AF772736

Anthracite Ghosts

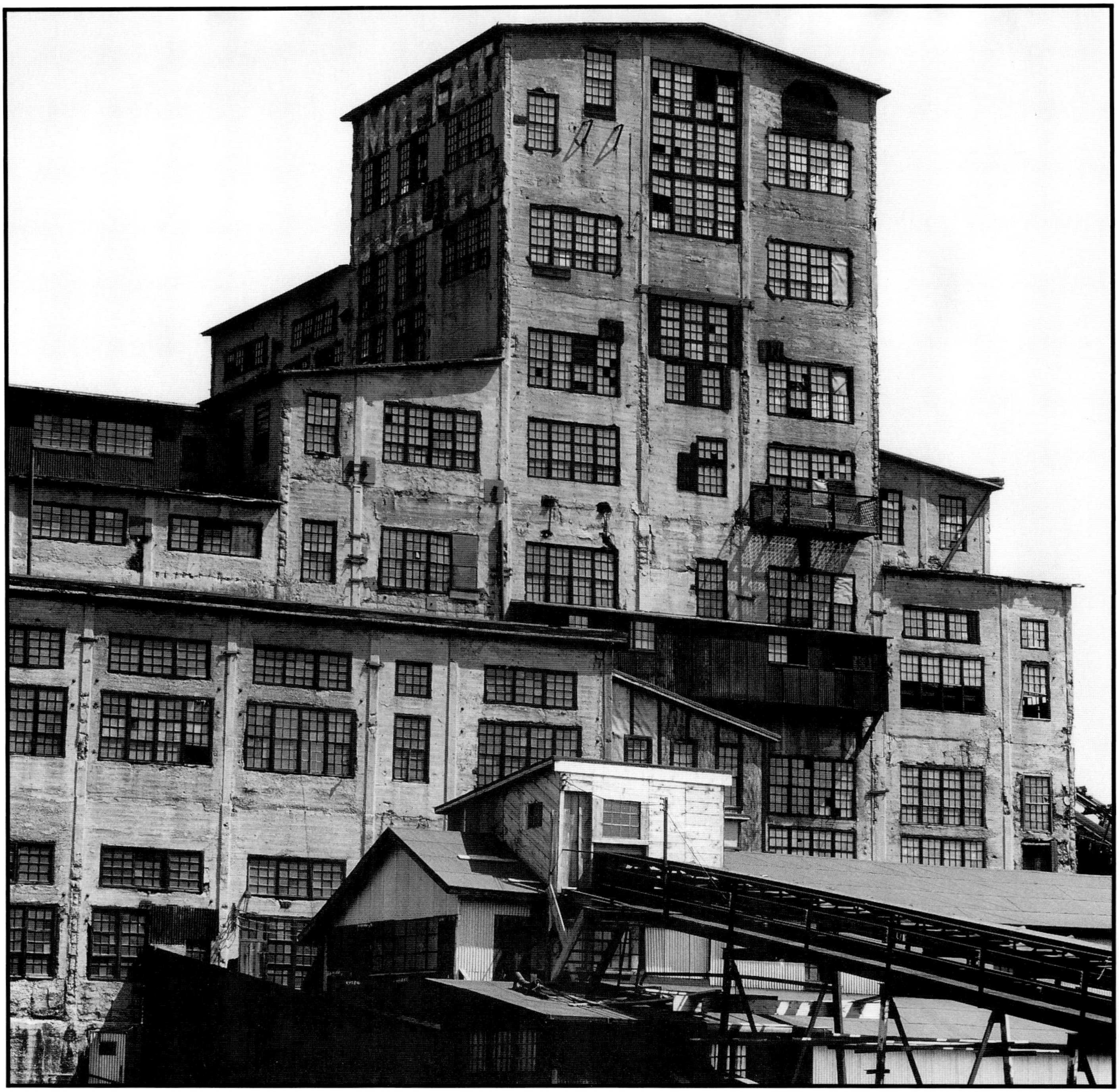

Anthracite Ghosts

Walter L. Dinteman

UNIVERSITY OF SCRANTON PRESS

1995

University of Scranton Press
Editorial Office:
Linden & Monroe
Scranton, PA 18510

ISBN 0-940866-44-7 (cloth)
ISBN 0-940866-45-5 (paper)

Marketing and Distribution:
Fordham University Press
University Box L
Bronx, NY 10458

Printed in the United States of America
by Thomson-Shore, Inc.

Composition by The Composing Room
of Michigan, Inc.

Designed by Laury A. Egan

Contents

Foreword

Anthracite Ghosts documents a photographer's romance with northeastern Pennsylvania's fast disappearing mine structures. Walter Dinteman, a veteran photographer, has spent several years roaming the area to film abandoned coal breakers and related buildings which are the crumbling monuments to another era. His photographs are not merely pictures of the buildings. Instead, he has moved in close to capture the texture of weathered wood, the pattern created by broken window panes, and the poignant mood called forth by long-stilled machinery.

"These old structures fascinated me because they lent themselves to photography," says Dinteman. "Then it seemed to me that as fast as I was photographing them, they were being torn down. I decided I had better hurry around and shoot as many as I could before it was too late. I wish I had come to this region earlier, because a great many of the buildings have disappeared in the past few years."

From Carbondale through the Lackawanna Valley to Taylor, on down the line to Wilkes-Barre, Hazelton, Eckley, Shenandoah, Shamokin, and way points, the photographer has walked among the ghosts of men long gone from the face and the tipple. "As you walk around these places," he says, "you find the ghosts everywhere. In the wash house, for instance, there'll be a pair of tattered trousers or a clothes basket with one shoe in it hanging from the wall. One thing that moved me was a stretcher, its canvas all rotten and tattered, leaning against a wall, a reminder of the high number of fatal accidents in coal mining. And if you go up high in the building, nine or ten floors above ground, you can hear the ghosts in the wind that whistles through broken windows."

In making his pictorial record, Dinteman has come to be deeply involved with his ghosts. "When I am around these buildings," he says, "I think of the things that went on: the hard lives of the men and their families, the stables of mules spending their lives underground, the legend of the Molly Maguires, the strikes and strikebreakers and the people really getting their heads bashed in. But I am not depressed. The whole scene is nostalgic and I become excited because it's as though I were walking back into an almost forgotten time. After I've gone to these places and have spent three or four solitary hours roaming around it, I feel as though I have some share in its history, as though I had lost a friend."

Does Dinteman feel any kinship with John Horgan, the well-known photographer of the coal mining industry?

"In a way, yes," he replies. "Horgan and I both roamed through the mountains of northeastern Pennsylvania, photographing the anthracite coal scene. However, he lived between 1840 and 1924. He photographed the collieries when they were alive, at the peak of coal production. I photographed them when they were ghosts. His photographs represent the beginning of the Industrial Revolution, whereas mine are at the end of that period."

Mr. Dinteman has taken hundreds of photographs during his two years of week-end excursions through the back roads of the region. His system was to single out an area and drive around in it. When he sighted a culm dump, he would start looking around for the buildings associated with it. If he was lucky, he had a new subject.

Scott Thurston

Anthracite Ghosts

Introduction

What is anthracite? Where does it come from? How does it relate to ghosts? The answers to these questions lead to the history of anthracite country. The anthracite industry flourished and then declined, leaving decaying structures isolated in quiet mountain valleys where wind whistles through timbers and rattles loose sheet metal—the environment of ghosts.

Anthracite is the hardest, purest, and hottest burning of the various types of coal. Formed from plants that grew approximately 250 million years ago, compressed through the ages, anthracite is 86 to 98 percent pure carbon and burns with a fierce heat. It can be made to burn cleanly and sulphur-free and can be converted to a gaseous or liquid fuel. "Coal gas" was used for street and home lights a hundred years ago. During World War II, Germany ran the machines of war with a liquid fuel derived from coal. With an all-out effort, coal could once again become a prime energy source.

The major anthracite deposits lie in a concentrated area of approximately 500 square miles in northeastern Pennsylvania and are found in layered seams down to a depth of a thousand feet or more.

Anthracite coal was discovered by accident in Colonial America by a group of New Englanders traveling through Pennsylvania in 1762. They built a fireplace of black rocks and were amazed to see this fireplace burst into flames and burn with an intense heat. This would have been no surprise to the Indians, who had had a knowledge of coal for a long time, calling it "black burn rocks."

Around 1769 anthracite—hard coal (bituminous is soft coal)—was known to exist near Wilkes-Barre, Pennsylvania, where a blacksmith, Obediah Gore, used it in his forge. Eight years later, the first shipment of coal was made from Wyoming Valley down the Susquehanna River to Carlisle, Pennsylvania, to be used in the manufacture of firearms.

Commercial development was slow. Little more than a ton a day was being produced by 1820. But during the U.S. Industrial Revolution hard coal replaced wood as a conventional fuel, and by 1865 nearly 12 million tons were being produced yearly. Anthracite's Golden Age was the decade 1910–1920. Peak employment in 1914 was 180,000. Production peaked in 1917 at 100 million tons. During the next fifty years, coal was gradually displaced as cheaper oil made inroads into the market. In 1930, 151,000 employees produced 69 million tons of anthracite. Only 7 million tons were produced in 1970.

At one time there were 315 large anthracite mines in Pennsylvania. In 1970 there were only 70 small mines operating.

The decline in anthracite production resulted not only from competition from oil and natural gas. A long history of labor disputes has taken its toll. Beginning in the nineteenth century, exploitive conditions led to strikes and bloodshed. Corporate owners, such as Andrew Carnegie, Andrew Mellon, and J. P. Morgan, were despised by the miners, who demanded an end to conditions that paid children only pennies for long, grimy hours in the mines. Coal barons hired detectives, professional thugs, and state militia to break strikes with clubs and bullets. Pennsylvania miners, claiming an eye for an eye, formed the Molly Maguires and engaged in reprisals against the bosses and armies of company police. The Molly Maguires were blamed for 42 murders, 162 felonious assaults, and myriad other acts of violence against coal companies and the railroads. They were accused of destroying, by fire, 30 coal breakers in Northumberland County between 1860 and 1909. Schuylkill and Carbon Counties suffered even more.

The militant Molly Maguires are history. Today a miner's wages and working conditions depend on the give and take at the contract-negotiating table. An excellent movie, *The Molly Maguires,* was filmed around Eckley in 1969, where an early wooden breaker was rebuilt and remains today as a tourist attraction.

Anthracite has the highest average cost per unit of all the fossil fuels. This is due largely to unfavorable geological conditions which present physical difficulties in mining steeply pitching coal beds. The deepest mines were about 2,200 feet below the surface. Some of the remaining coal is at a depth of 4,000 feet or more. Most of the anthracite is below water level; thus, another factor in the high cost of mining anthracite: mine shafts penetrating below water level require constant pumping. As the market shrank, mines closed and pumping ceased. Each time a mine was abandoned and allowed to flood, the next mine down the slope had an increased amount of water to pump. Too much pumping would become too costly and the mine in turn would be closed. Finally, all the mines below water level were flooded. The final blow came to the Northern Coal Field in 1959 when mining too close to the bottom of the Susquehanna River caused the mine roof to collapse and the river poured into the mines. Today, all the existing mines are above water level. One large mine still in operation is the Kocker Coal Company's mine in Valley View, Pa. Its main shaft is bored almost horizontally into the mountainside; thus, surface water seeping into the mine flows out by gravity.

The environmental problems associated with mining are considerable. Acid mine drainage contaminates water resources and presents a costly problem in water purification. Abandoned mine cave-ins cause land subsidence. In Taylor, Pa., in 1970, whole sections of the city were caving in. Culm bank refuse was pumped into the mines under the streets of Scranton to avoid the same fate there. Mine fires which burn for years destroy valuable reserves. Smoldering culm banks create noxious fumes, creating a threat to public health and safety and blighting scenic views.

Culm banks are mountain-high waste materials from mines, intermixed with coal lumps too small to sell. Sometimes these culm banks catch fire by spontaneous combustion. Two burning culm banks near Scranton have recently been extinguished by leveling the banks. They had been burning for more than 20 years. In 1970 there were 30 anthracite banks and mines burning in the state. Attempts to extinguish such fires have often failed; some have been burning for decades.

Today, with the high cost of oil, the price differential with coal may diminish. Perhaps the landscape will soon be dotted with processing plants converting coal to a liquid fuel. But will these plants appear in anthracite country or in areas where bituminous coal is easier and cheaper to mine? While we ponder the future, there remain many abandoned mines and ghostly structures of former collieries and breakers in northeastern Pennsylvania. Viewing these decaying ruins, isolated in their tranquil mountain surroundings, one can easily visualize the activities in a not too distant past, of the hard lives of the men and their families, the stables of mules spending their entire lives underground, the legend of the Molly Maguires, the strikebreakers, and people really getting their heads bashed in.

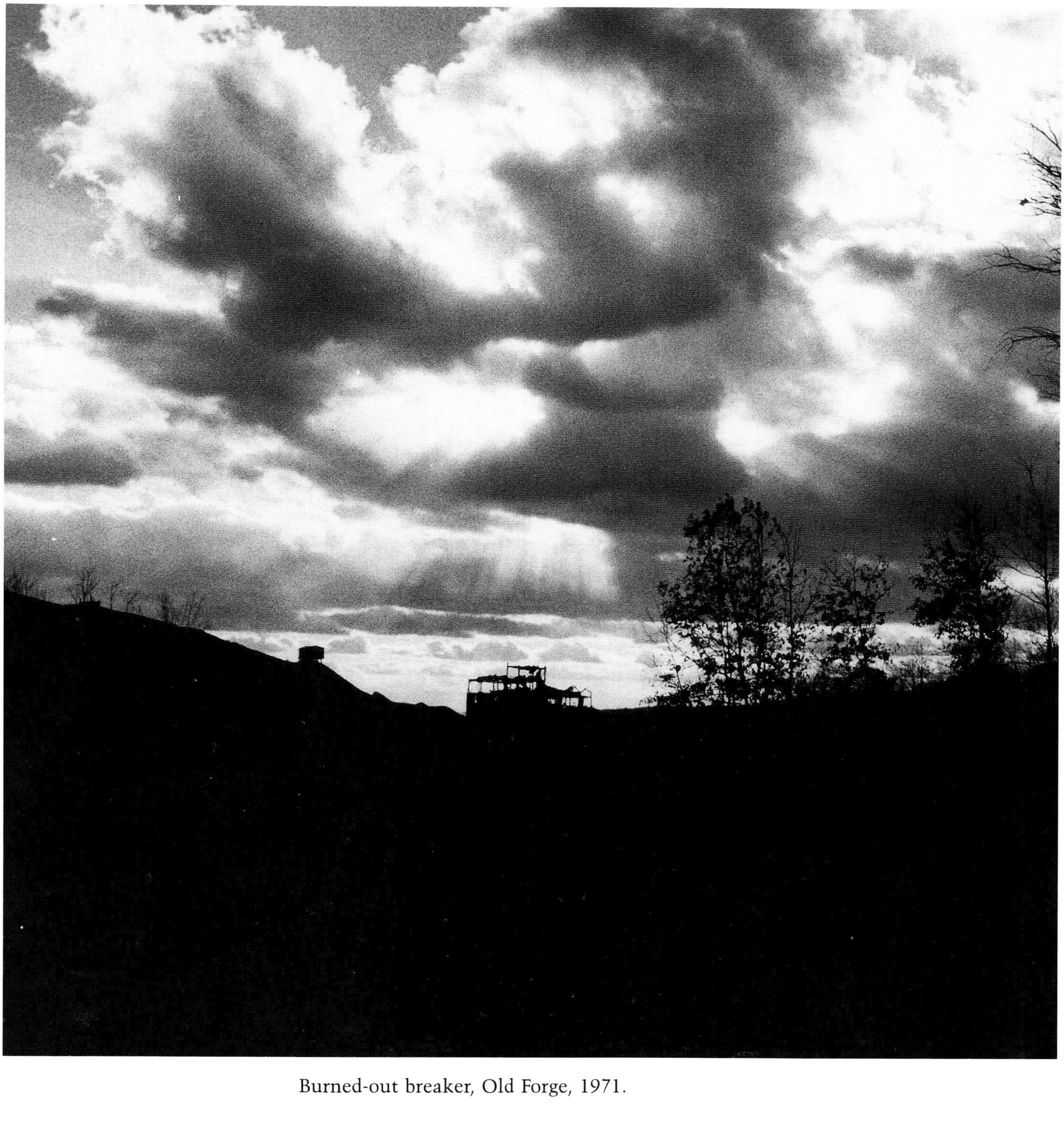

Burned-out breaker, Old Forge, 1971.

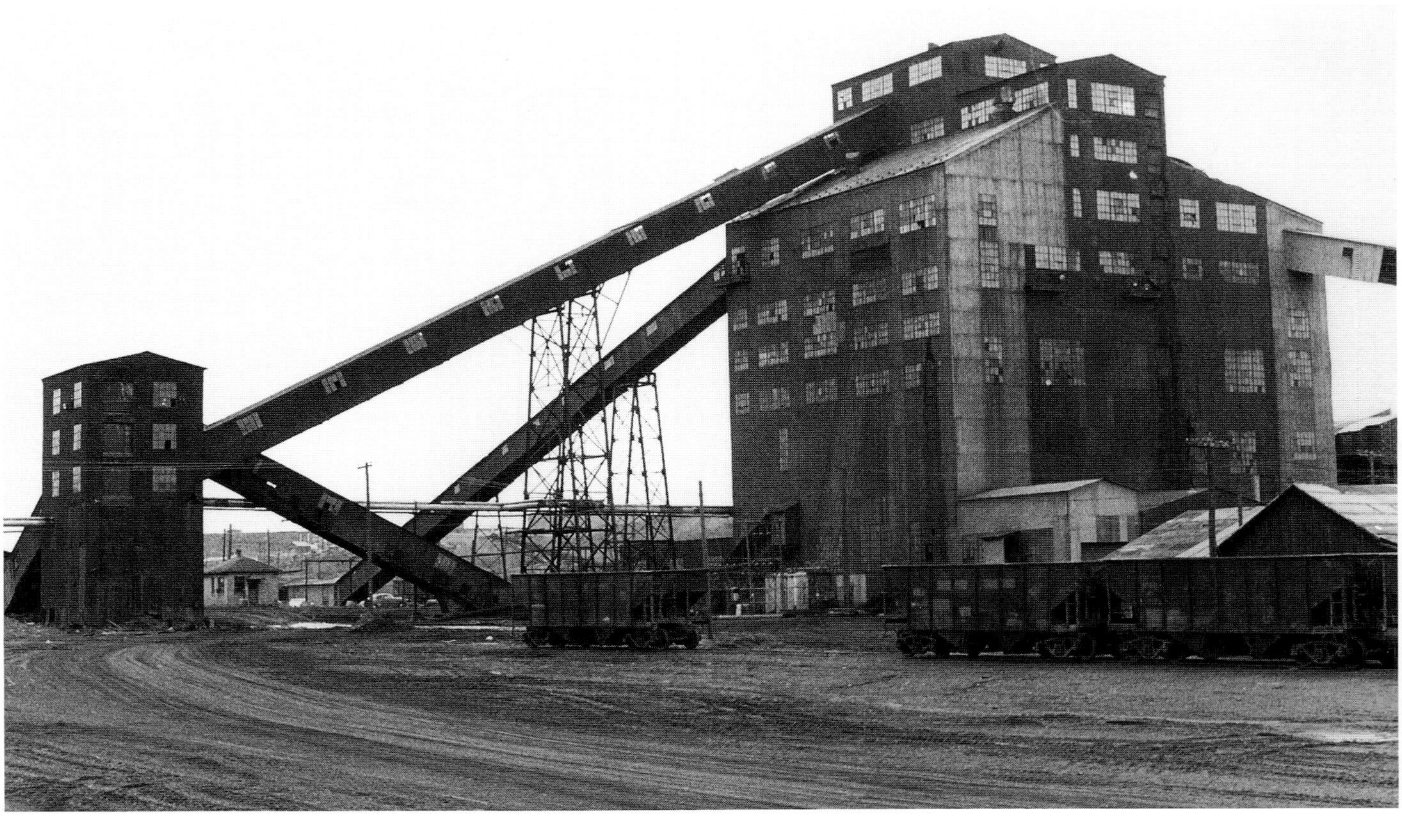

COAL BREAKERS: a straggling hulk of a building, more than a hundred feet tall, an imposing sight against the barren landscape. These vast factories for processing coal were the most characteristic feature of the anthracite country. Most stand silent now, rotting remnants of a once dominant industry.

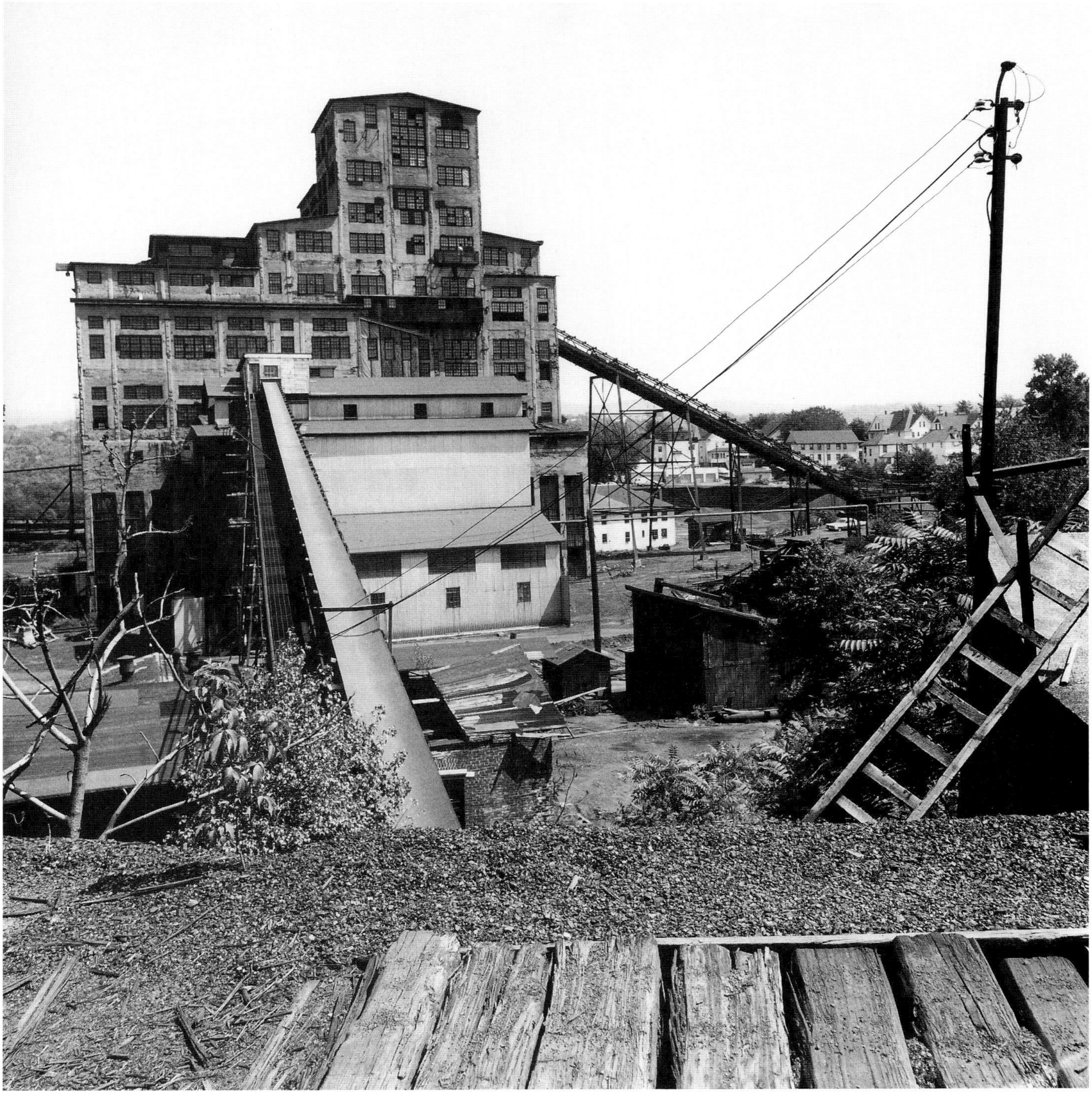

Moffet Coal Colliery, Taylor, 1971.

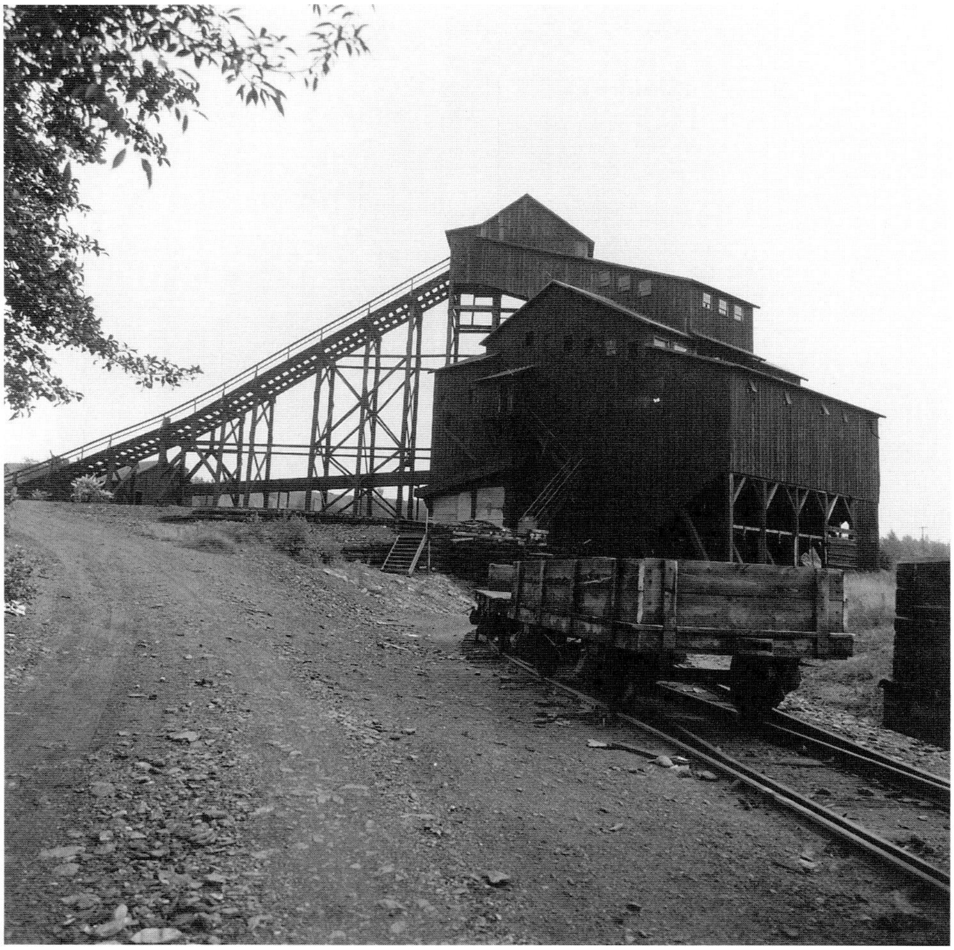

Replica breaker, circa 1900. Built in Eckley for filming scenes of the movie *The Molly Maguires.*

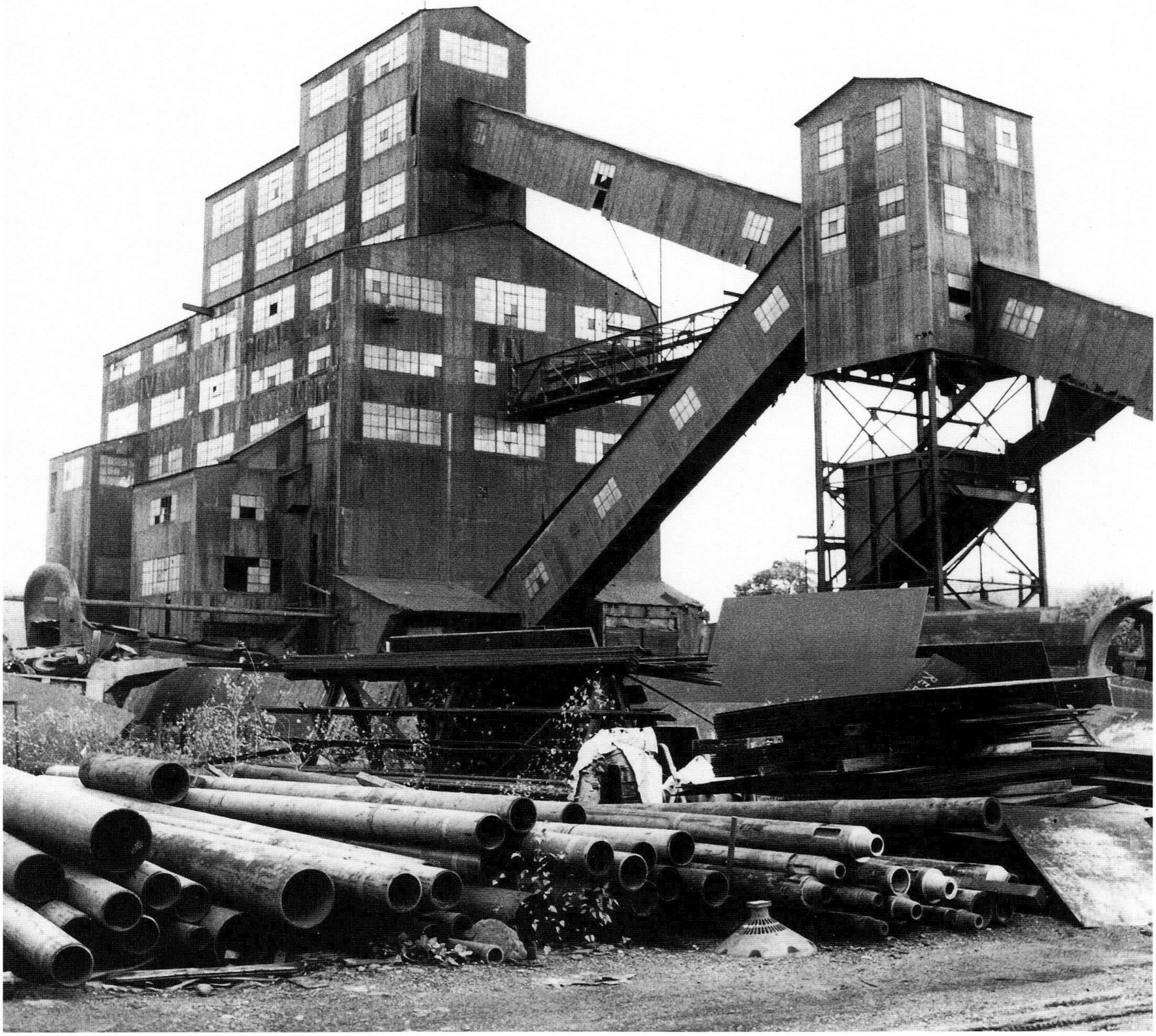

Breaker at West Pittston, 1971.

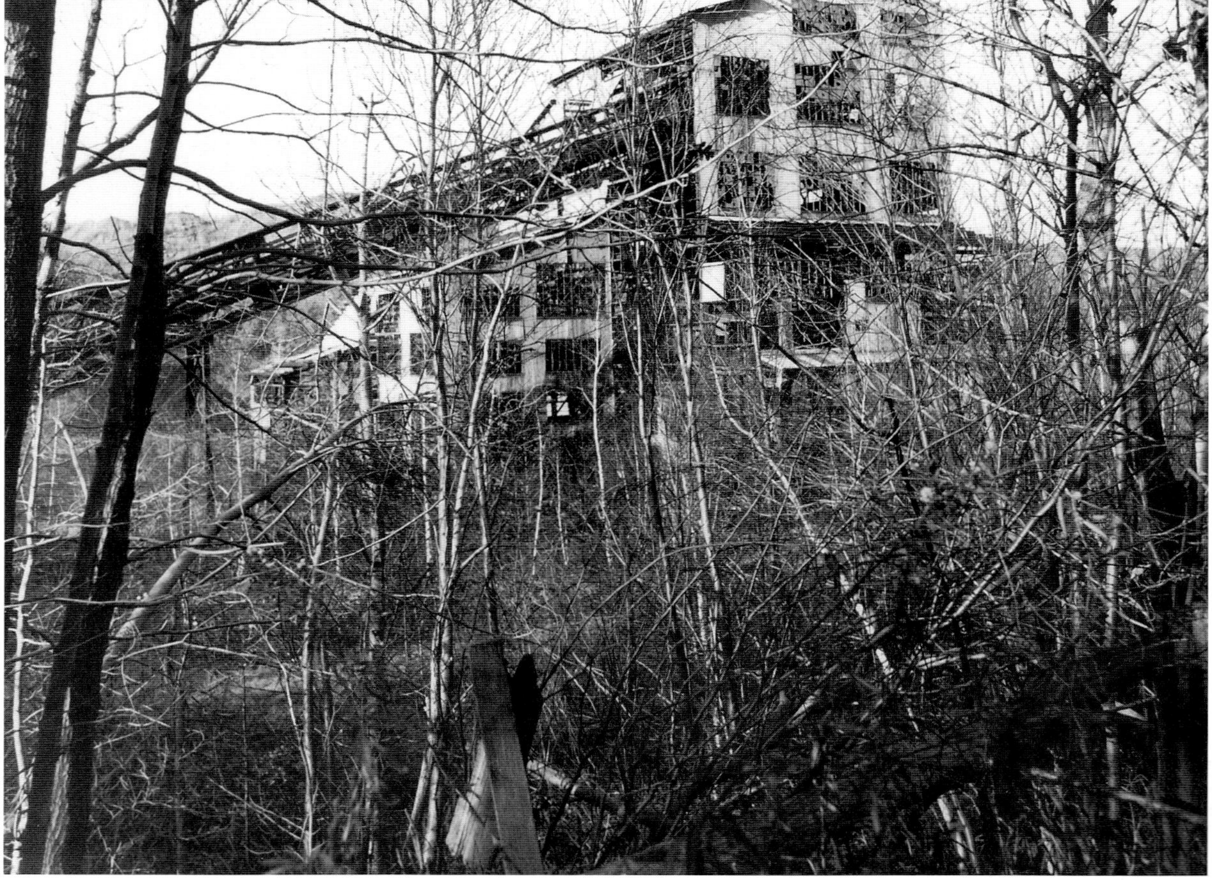

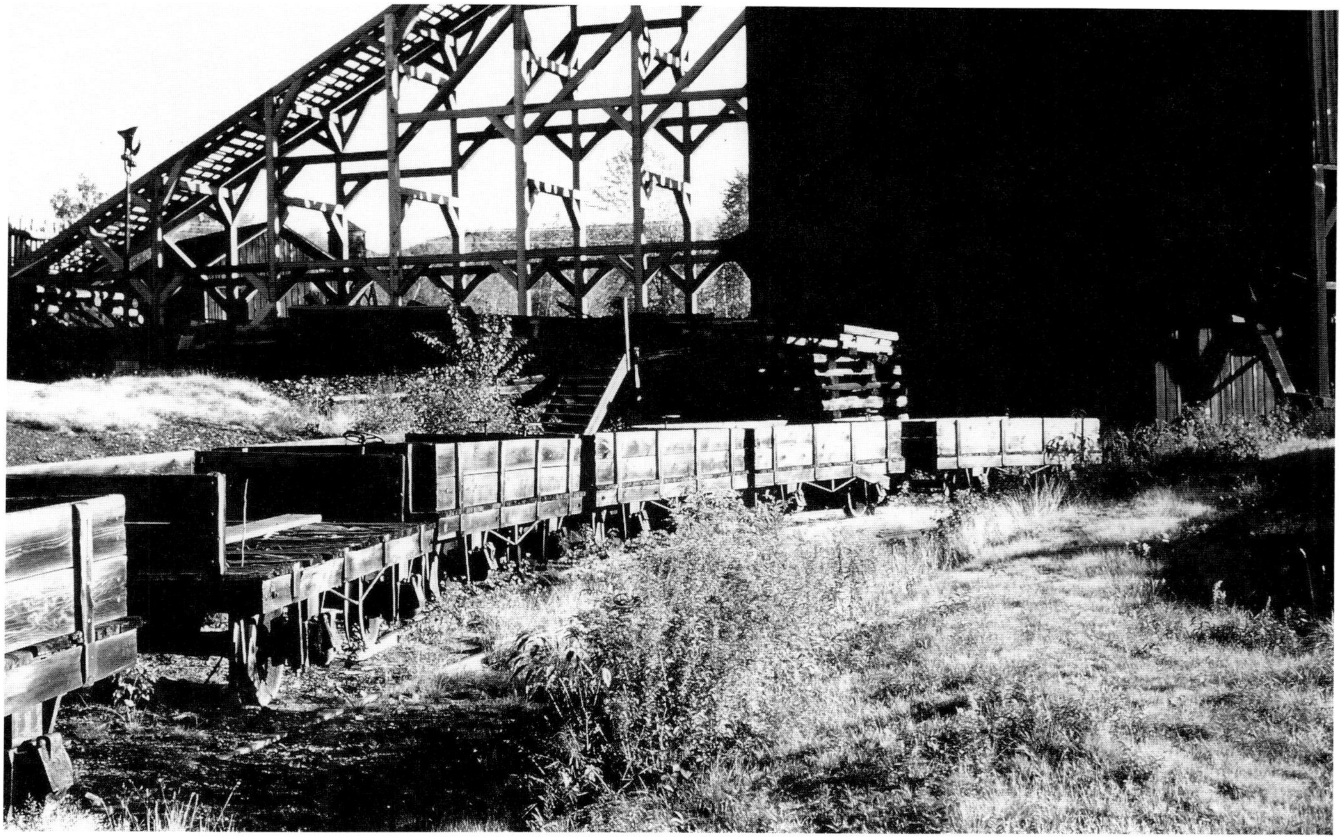

Coal cars at the Eckley Breaker.

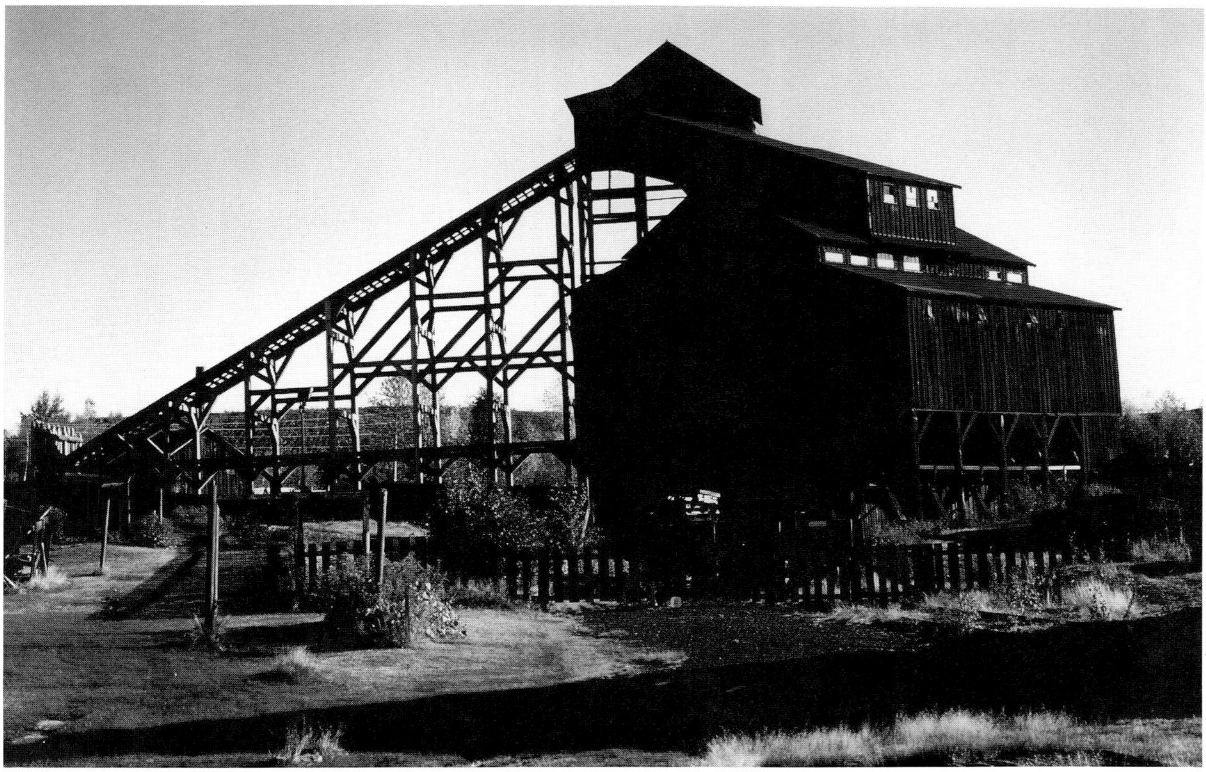

Eckley Breaker.

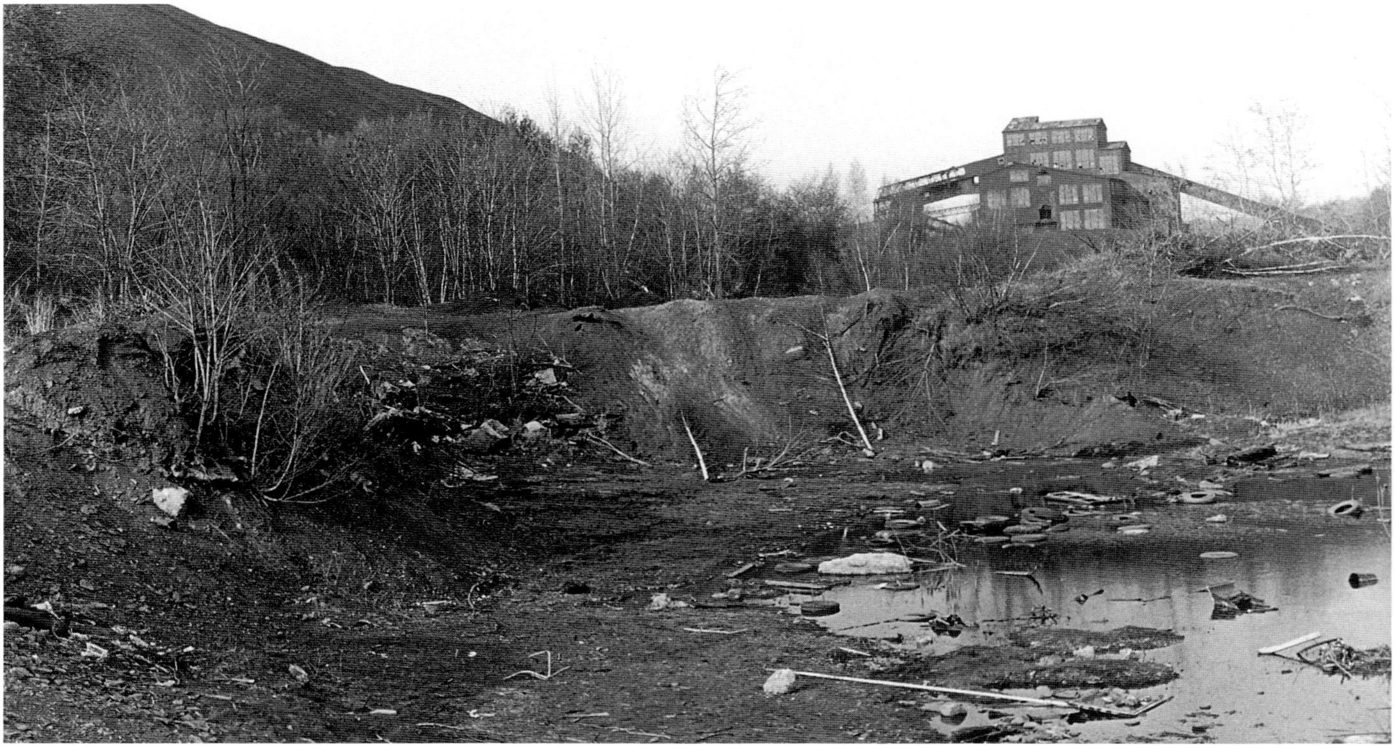

The breakers, equipped with the necessary machines, are expensive. They are usually built of wood and require constant repairs. The cost of opening a colliery, with its equipment, varies from $200,000 to $750,000, which must be returned by the time the coal is exhausted or the investment results in a loss, as the plants are worthless when all the coal has been mined.
—*Report to the President on the Anthracite Coal Strike,* 1902

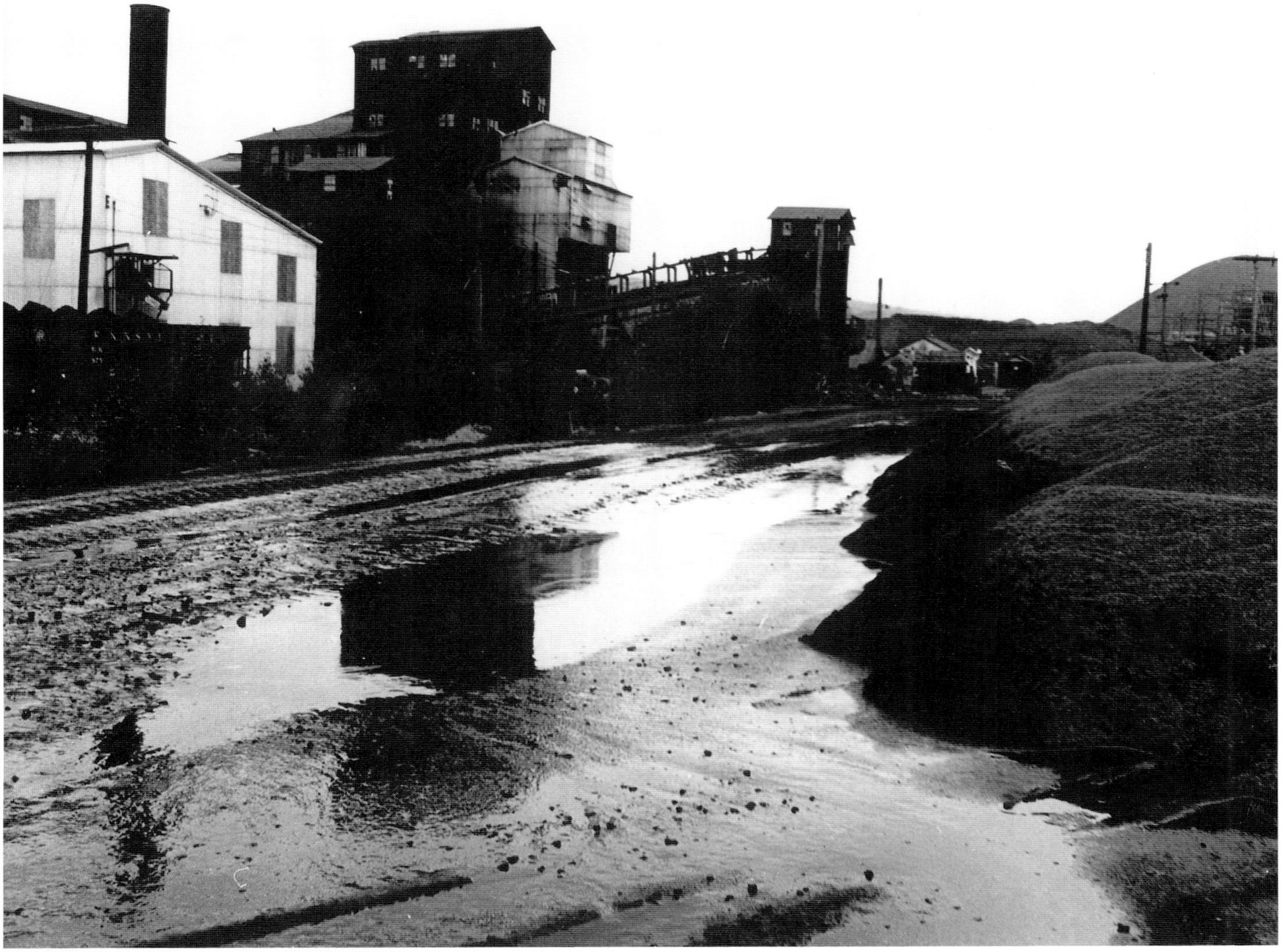

Black dust, water, and mud dominated the coal colliery environment.

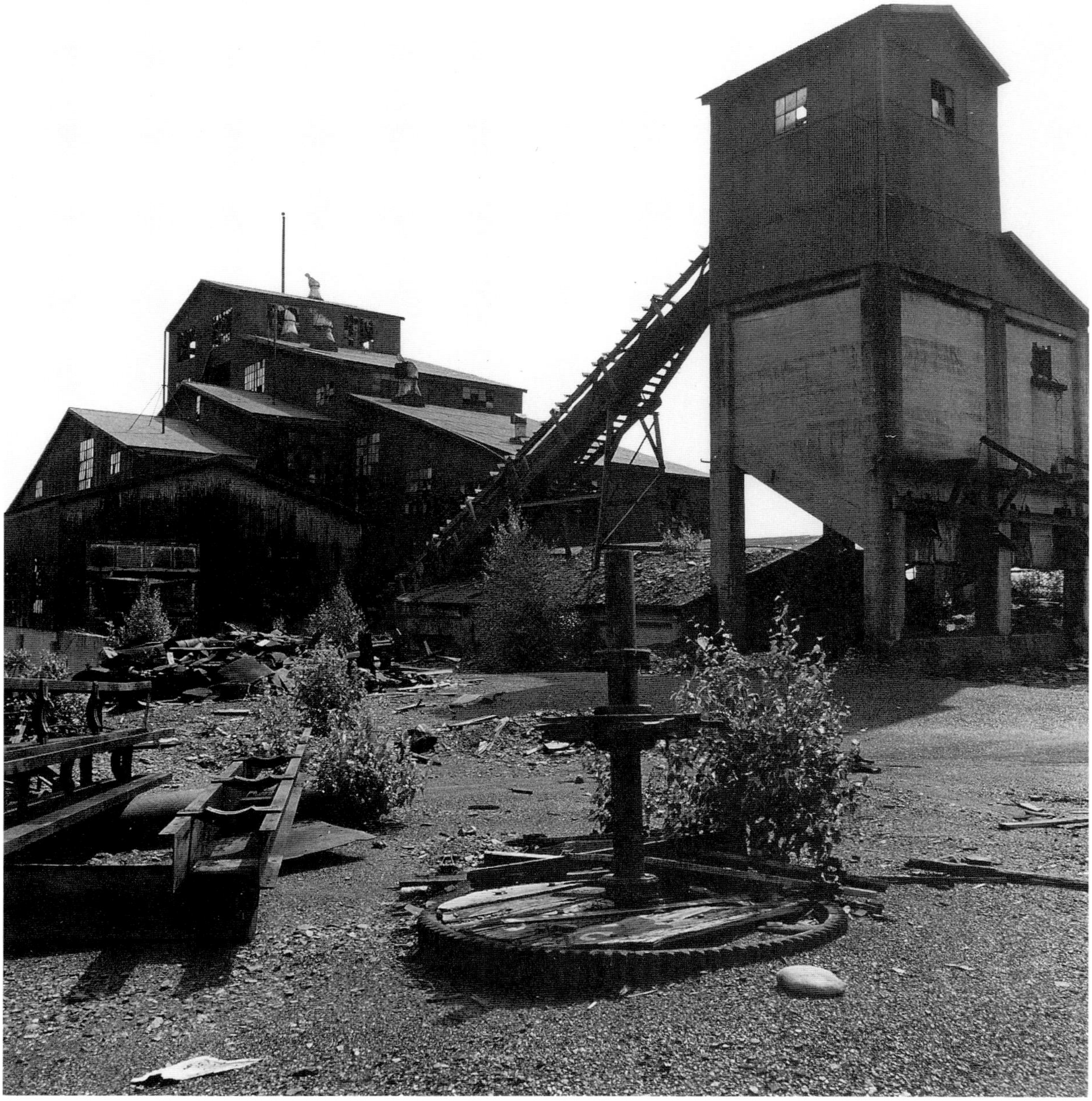

A working colliery at Dickson City, 1972.

Men and machines journeyed in and out of the ground like ants in an anthill. And the black coal dust lay over it all: men, buildings and machinery; its acrid smell was everywhere. In the distance, black culm piles loomed higher than the breaker; around them moved little yellow streams, poisoned by waste-filled waters of the mine, sluggishly licking at the barren gullies and rock refuse lining its bank.—Maude Morgan Thomas, *Singing in the Dark,* 1954

An act passed June 15, 1911, provided that all buildings inside any coal mine in Pennsylvania, including engine houses, pump houses, stables, etc., be constructed of incombustible materials. Enactment of this law came as the result of the Pancoast mine fire in Throup, April 7, 1911, when 73 persons lost their lives. The fire occurred in the engine house of the mine. It communicated the flames to nearby cars and through them spread to the tunnels and workings.
—Alexander Trachtenberg, *The History of Legislation for the Protection of Coal Miners in Pennsylvania,* 1874–1915

Later, cement construction was found to be unsatisfactory due to cracking from stresses which weakened the structure.

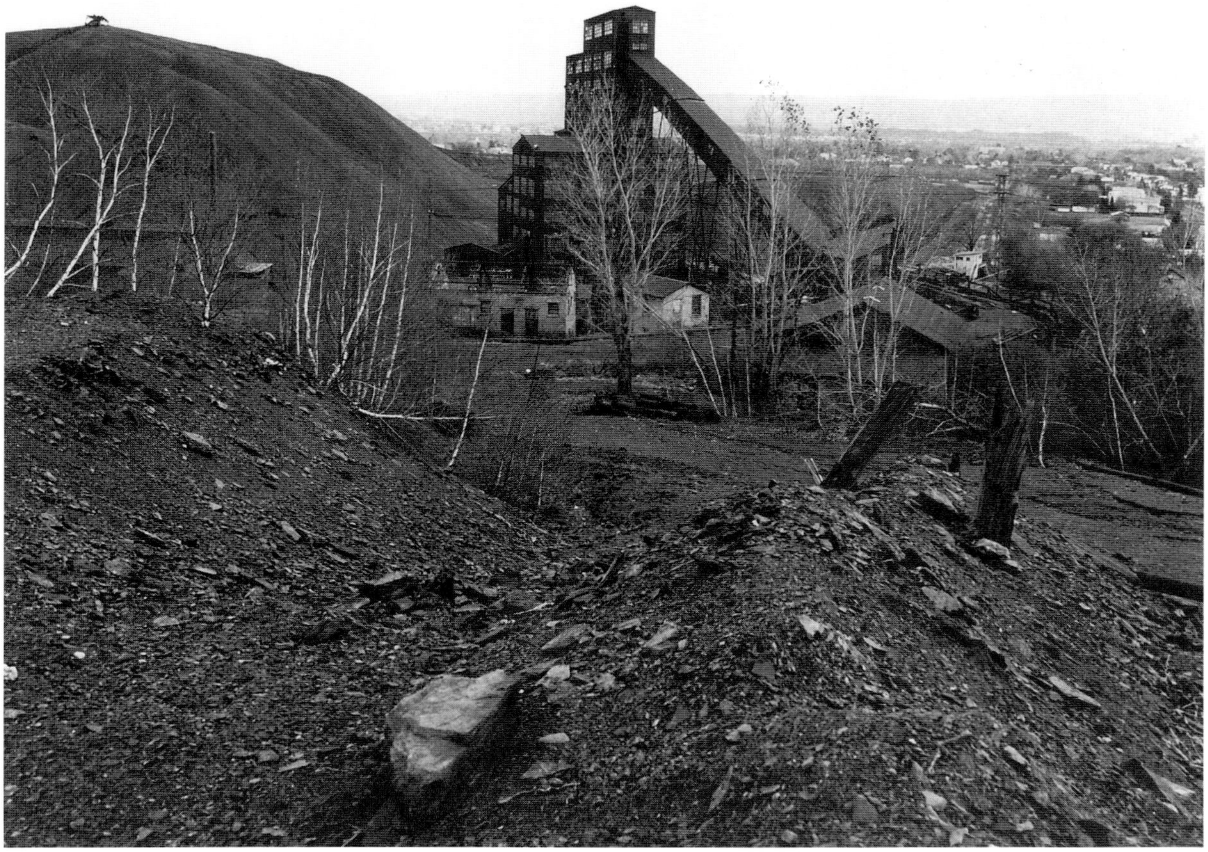

Heidelberg Breaker, Avoca; demolished in 1972.

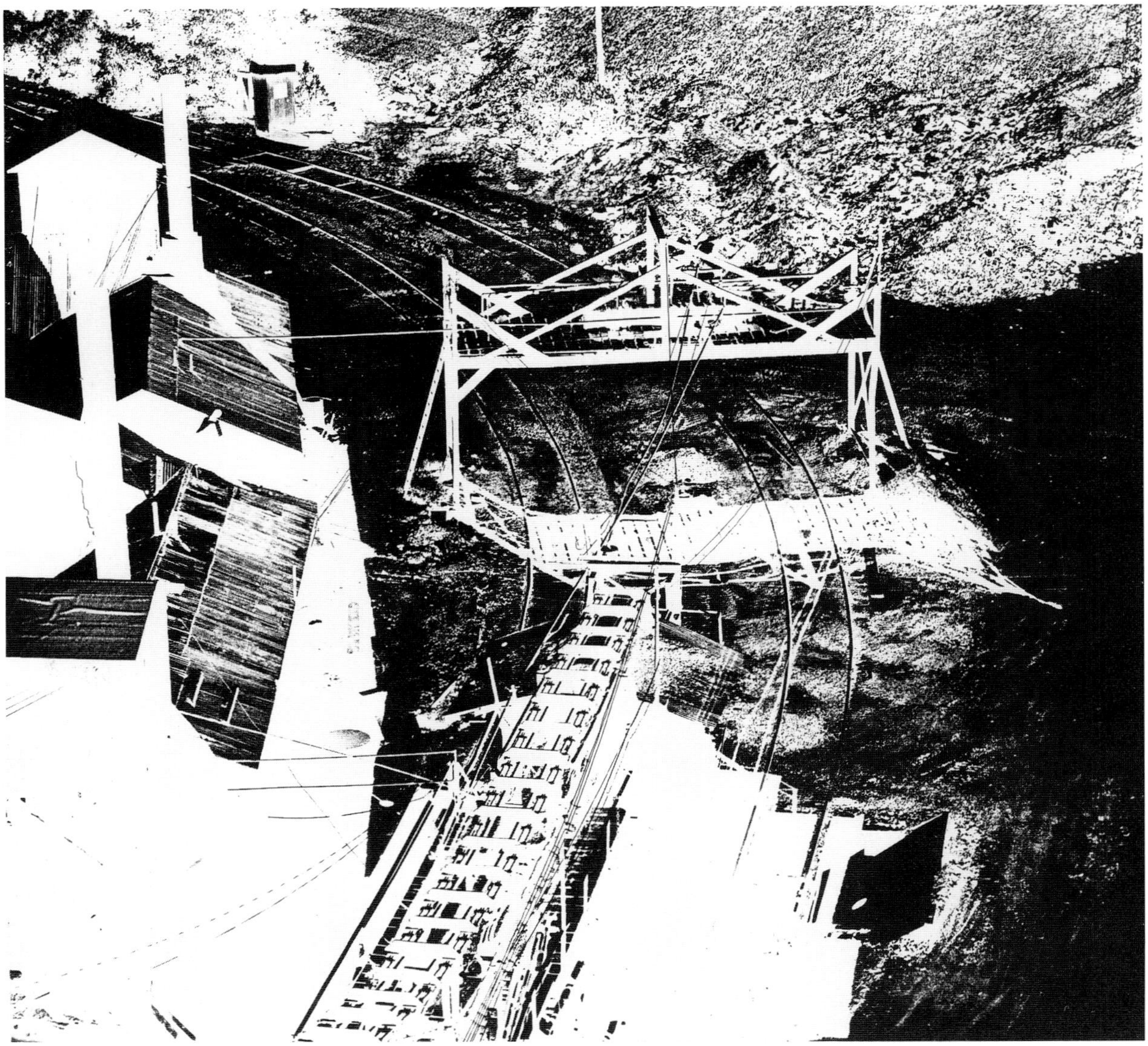

View from a breaker.

. . . shoveling, a stoppage of coal had piled up against the traps that open readily to release it into freight cars below. The work had been tiring and dangerous. Boys had been known to have lost their footing and to slide down with the sudden release of the coal. Some had been pulled out in time; a few had been smothered to death in the loaded freight car.
—Maude Morgan Thomas, *Singing in the Dark*

Taken from school ere his course is half ended,
    Sent to the breaker where danger is rife;
There, in its gloom, his young strength is expended:
    There is he dwarfed in the morning of life
Poor little Cracker Boy! Martyr to want and greed,
    Who thy sad destiny tearless can sing?
Parents at home, in the depths of their woe and needs
    Crave for the pittance thy labor can bring.
—Patrick F. Durkin, "The Cracker Boy," *Poems and Lyrics,* 1903

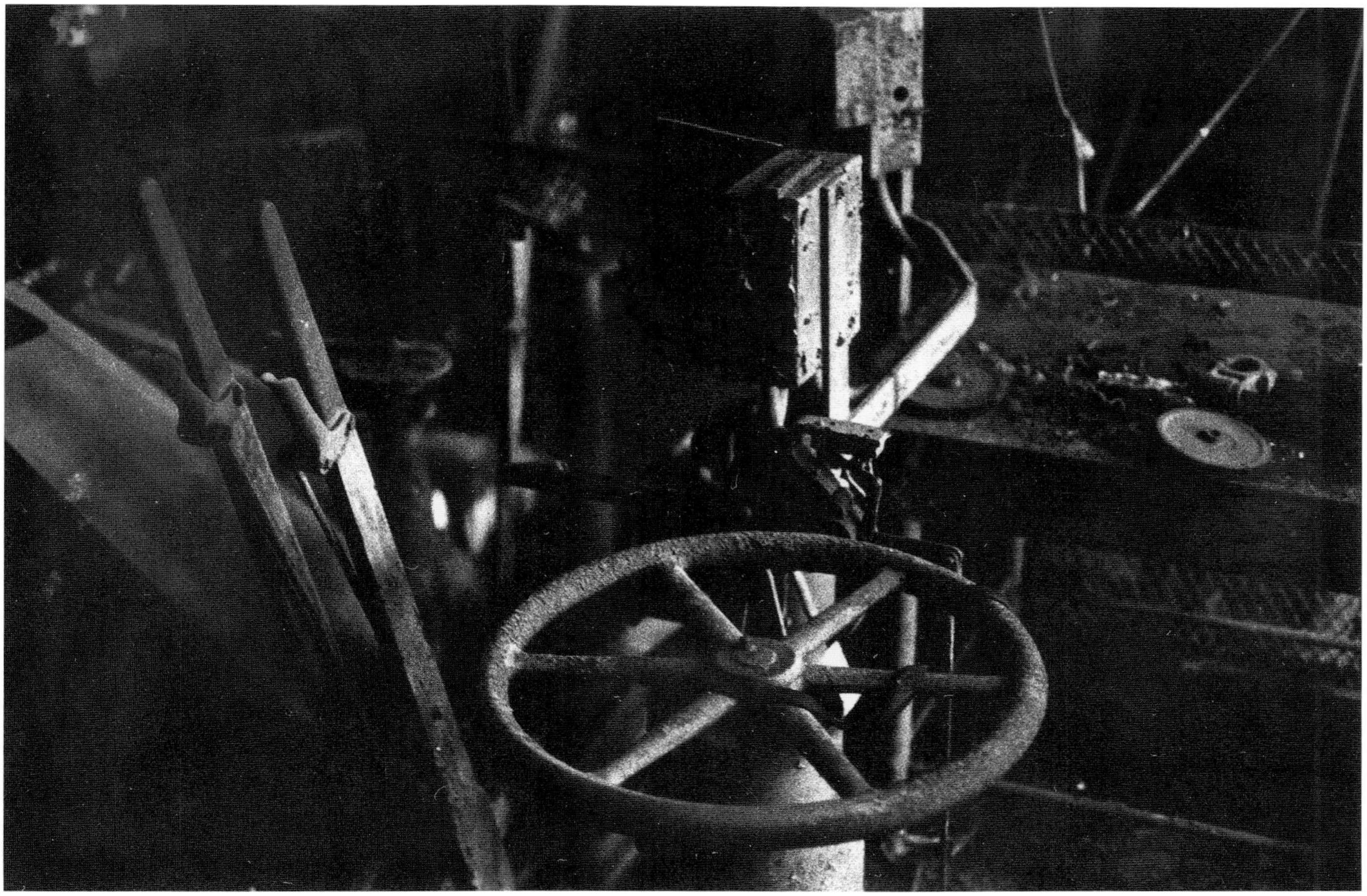

Gears and levers in an abandoned breaker.

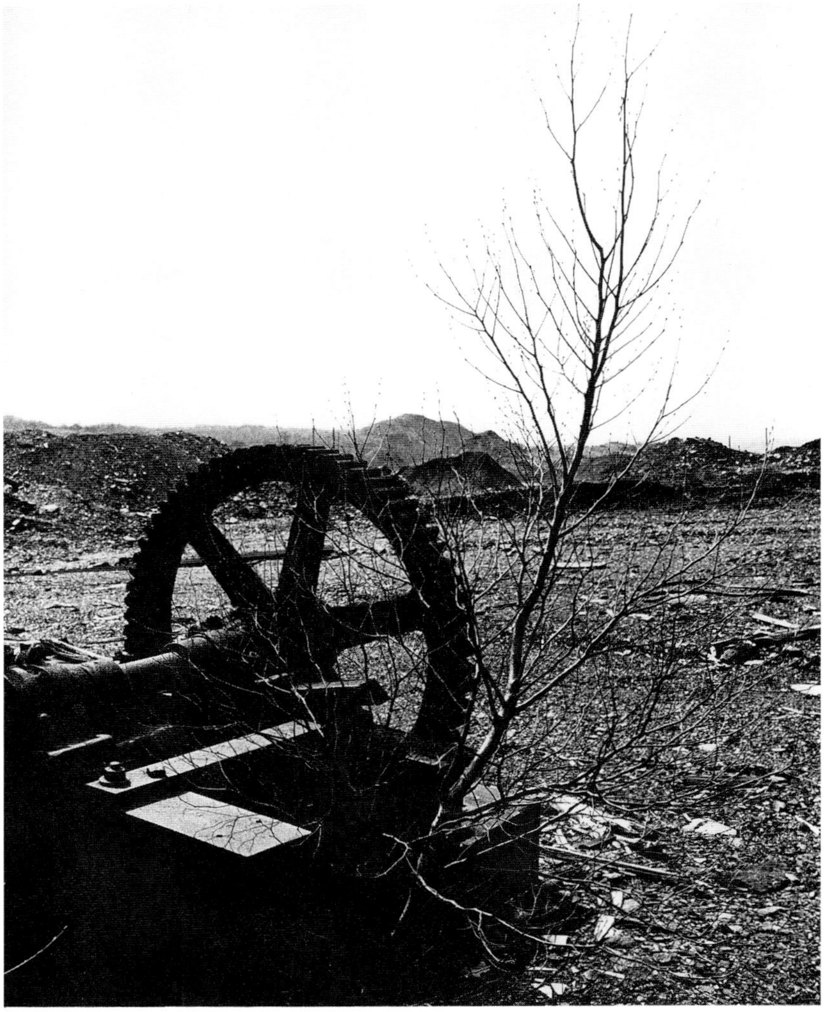

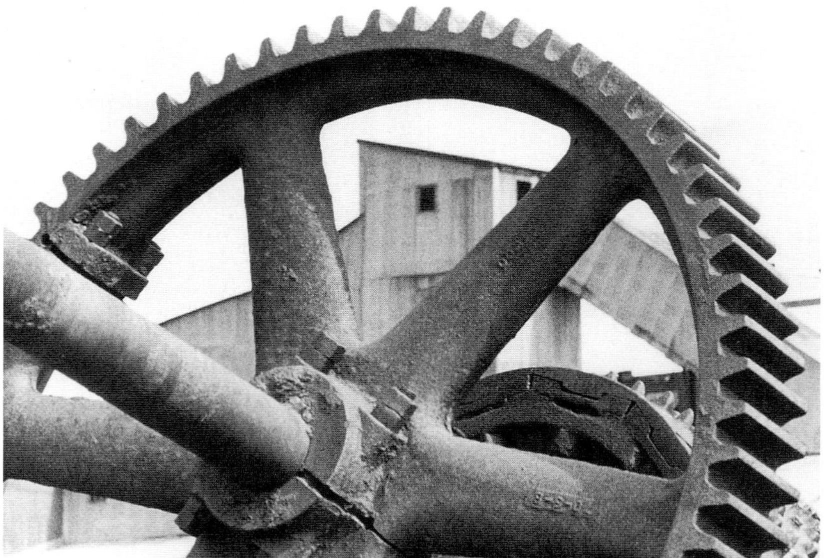

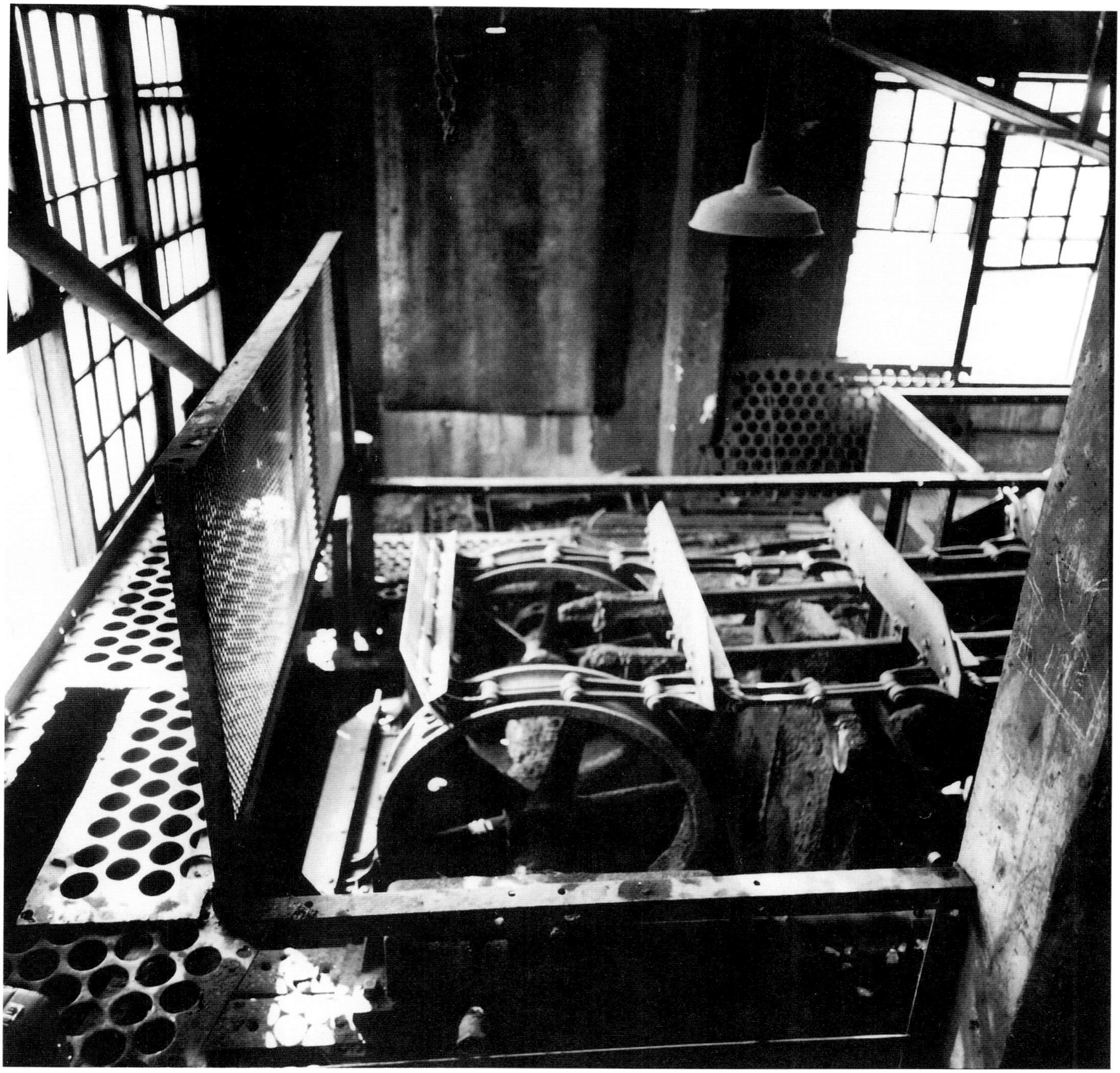

Machinery once used in elevating coal to the top of the breaker now rusts in an abandoned breaker. The floor grating is the same as used to divide different size coals.

The Anthracite Act of 1870 provided that no boys under 12 years could be taken to work in the mines. The age was raised to 14 years in 1885, allowing 12-year-olds to work outside mines. A 16-year-old limit was set in 1903.

"Did you like working in the mine? It sounds so hard." I looked at his still-adolescent shoulders and thought of the sweat pouring from the brow of the mighty Archie Konncheck.

"Well, when I first went to work five years ago I begun on the tipple and I sure hated that. I was a breaker boy and I had to stand there under that roarin' noise all day, so loud you couldn't hear yourself think. It near addled your wits. I had to pick slate outa the coal while it came down the chute. You git black from head to foot. Worse than diggin'."

—Laura Gilfillan, *I Went to Pit College,* 1934

"The dust is bad. It is dirty. You have never seen anything like it until you go down inside one of those holes. This type of work kills a man. I know I started in when I was nine years old at 45 cents a day. This coal mining, though, becomes a part of you. Despite the bad part, it is attractive to you. It just becomes a habit you can't live without it. It's sort of like smoking."

— Herman Lantz, *People of Coal Town,* 1958

"Well, that's what I liked about the coal mines was that on hot days like this, you were down there and you didn't have to suffer the heat. Well, I would say that is the main thing. You can go down there and work pretty hard. You get up a sweat while you are working but when you stop, you get chilly. A fellow could work and it wasn't hurting him too bad. Eventually they made the mines safer. They put up lots of safety lights and everything. Mining wasn't too bad then. It was pretty bad before that."—Joe Sleyzak, retired miner, died of "black lung" in 1979

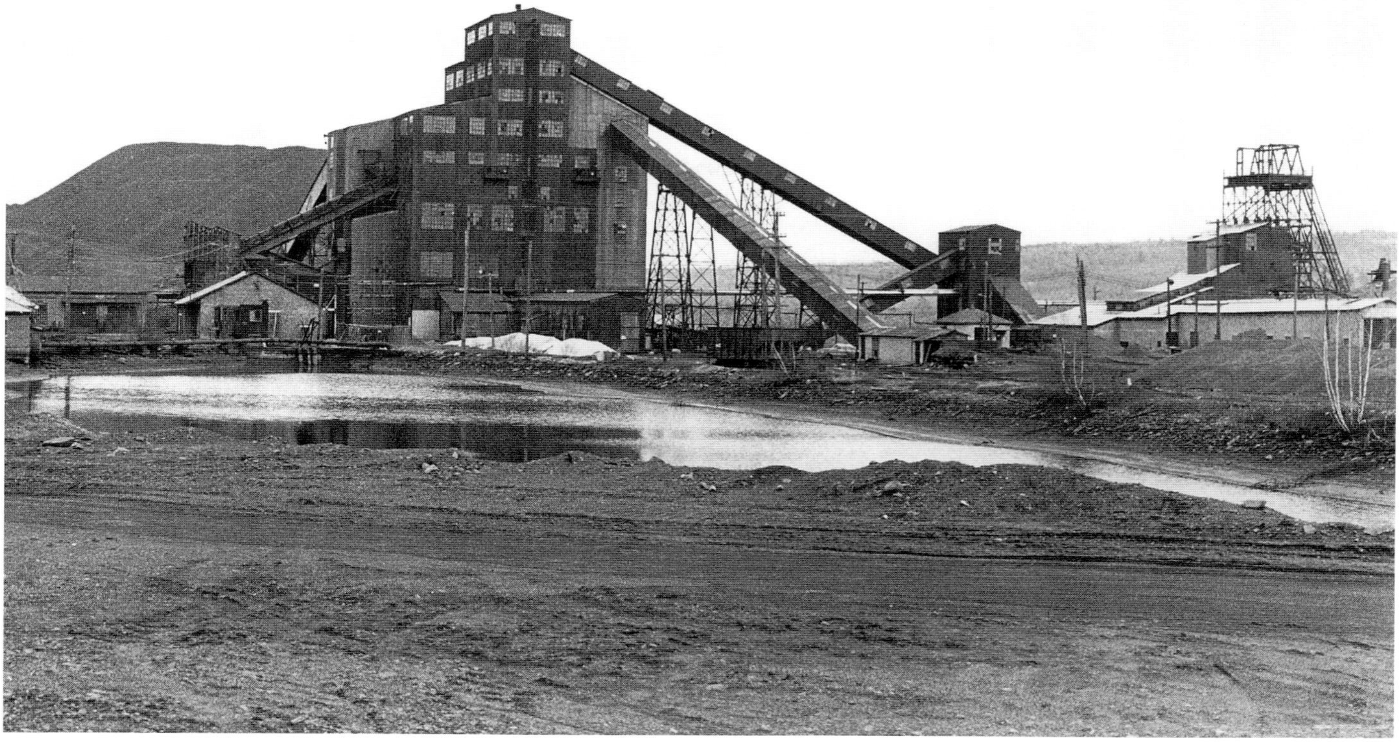

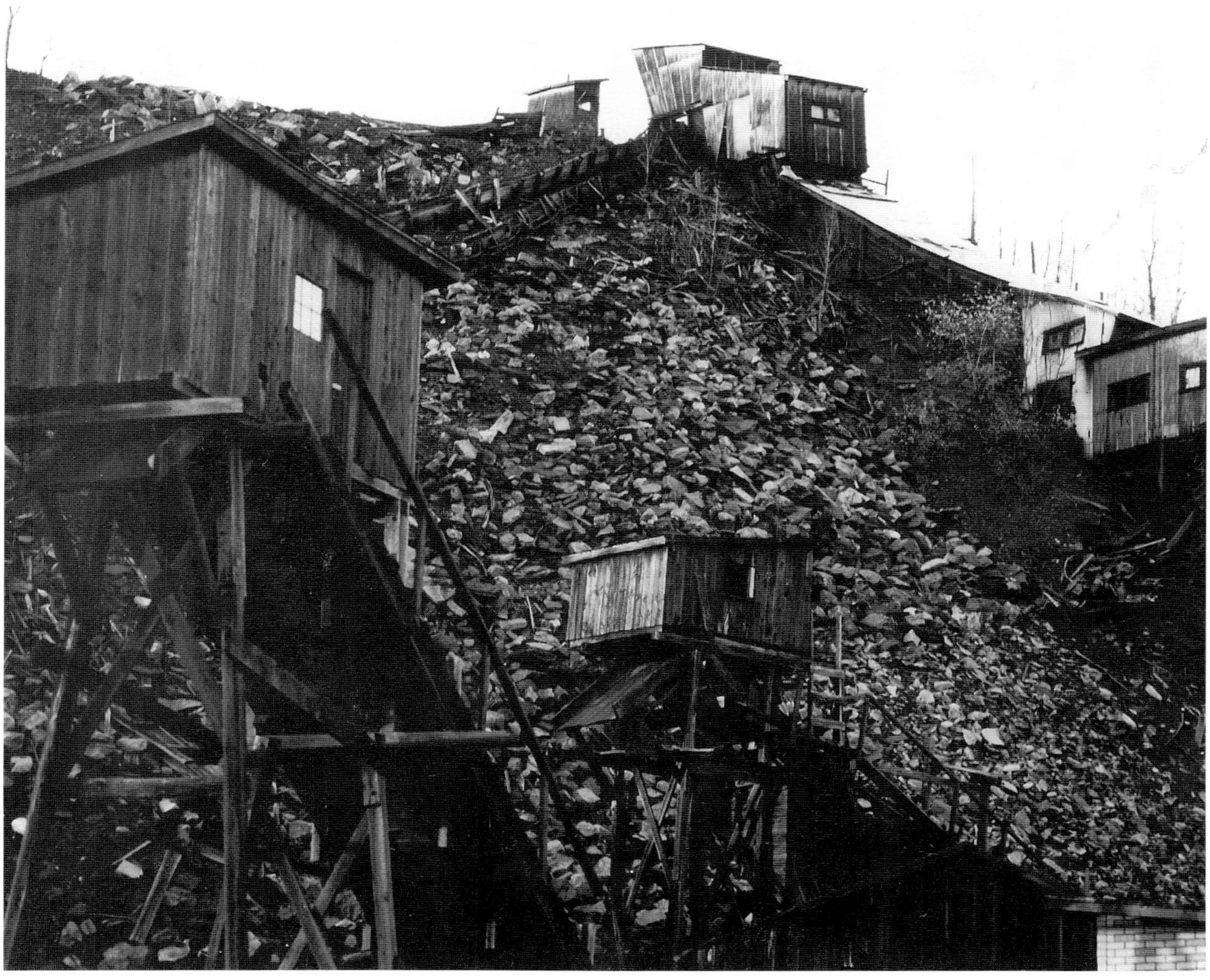

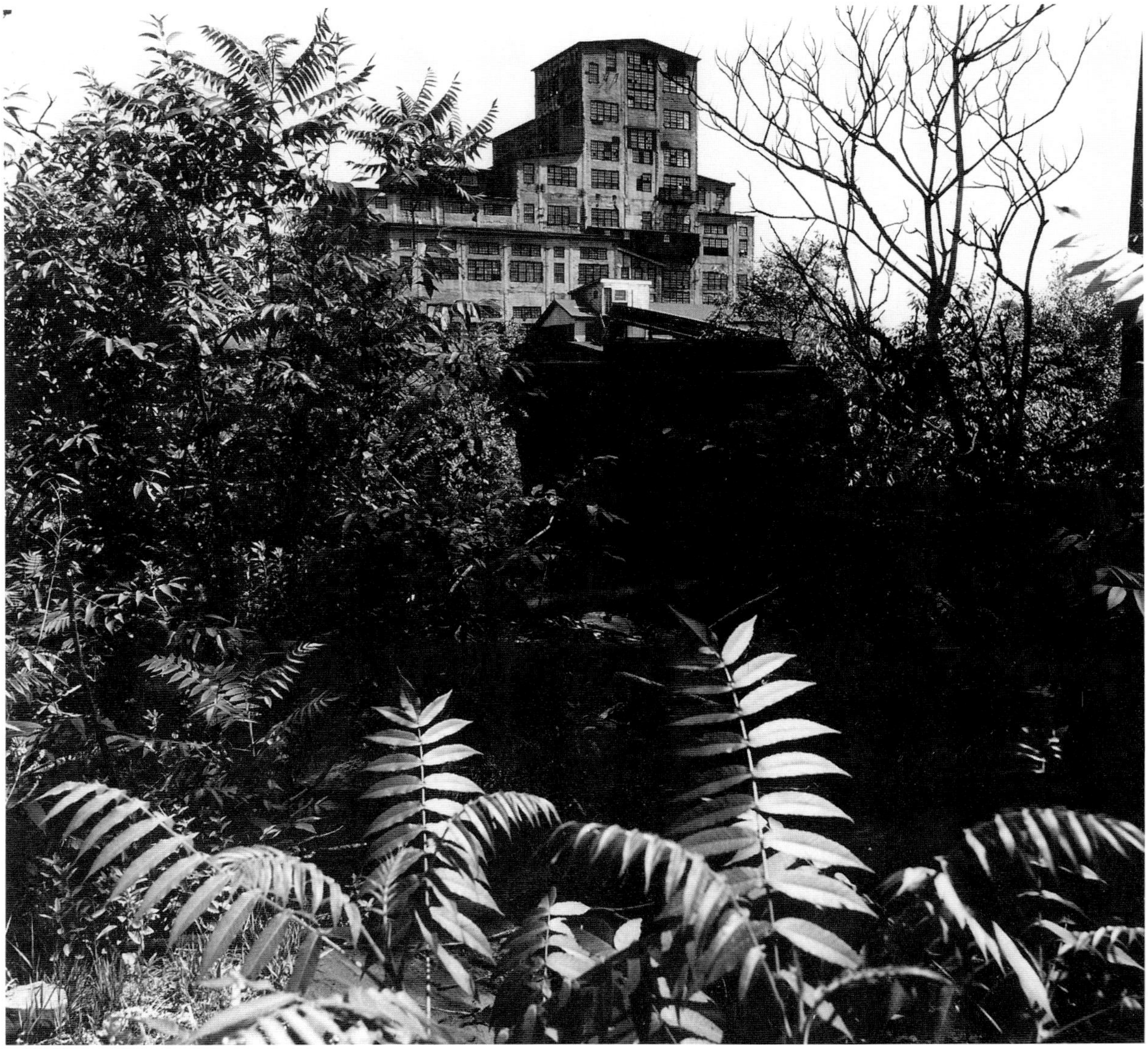

Weeds, brush, and trees taking over the Moffat Coal Breaker, Taylor.

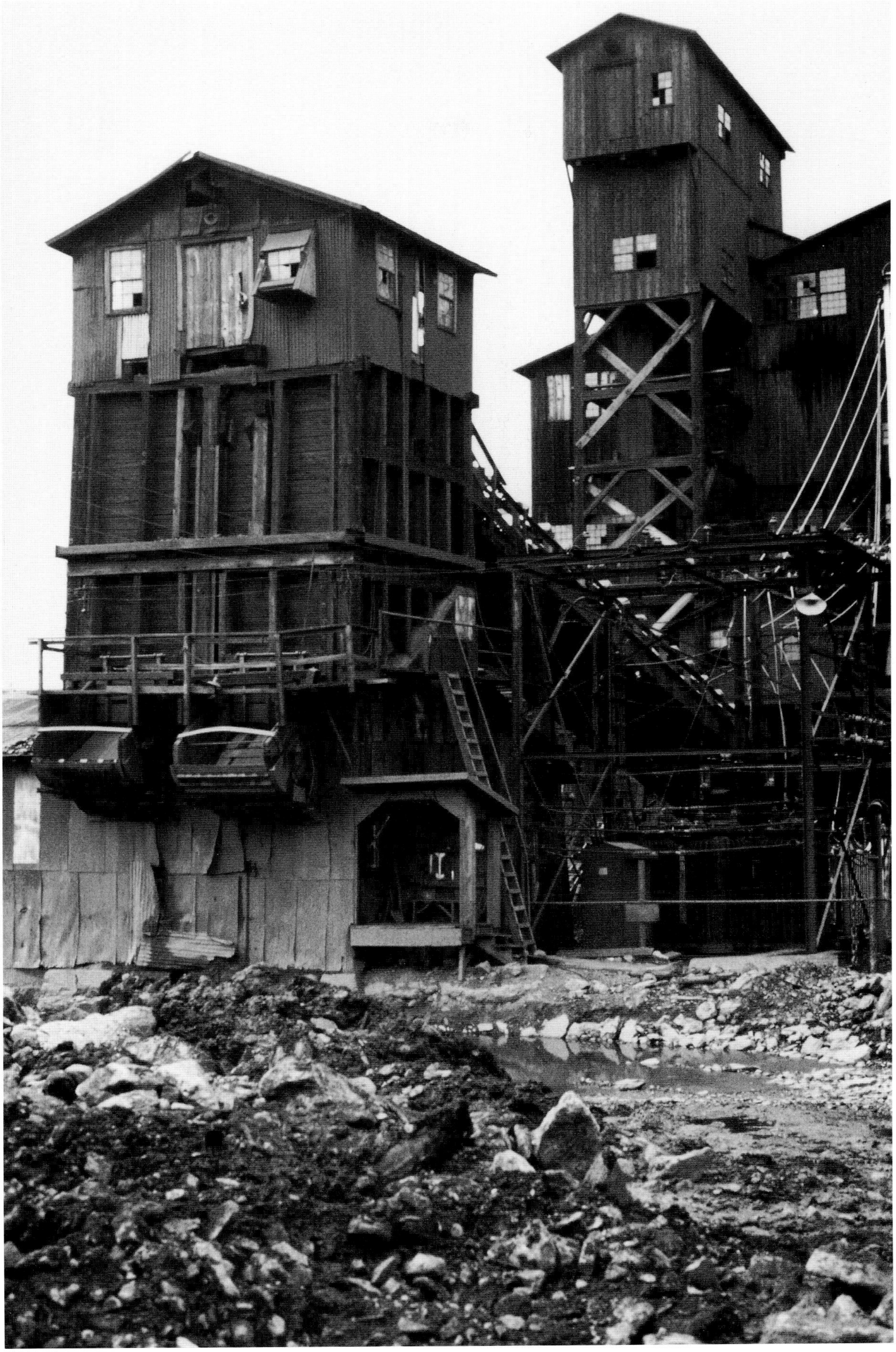

The first thing you will do boy, you'll pick up your drill,
You will then sound the top and your car you can fill.
If the top is bad, drill in a long hole,
Give it powder and tampin' and blow lots of coal;
You will then break the lumps, throw them in a car—
Wipe the sweat from your face—I tell you it's hard.
When your legs you must bend and your back you must stoop,
You will curse the first man who invented the scoop.
—Patrick J. O'Neal, 1878, *Songs and Ballads of the Anthracite Miner,* 1926

I'm getting old and feeble and I can work no more;
I have laid the rusty mining tools away—
For forty years and over I have toiled about the mines,
But now I'm getting feeble, old and gray.

I started on the breaker and went back to it again,
But now my work is finished for all time;
The only place that's left me is the almshouse for a home,
That's where I'll lay this weary head of mine.
—Anthony Bimba, *The Molly Maguires,* 1932

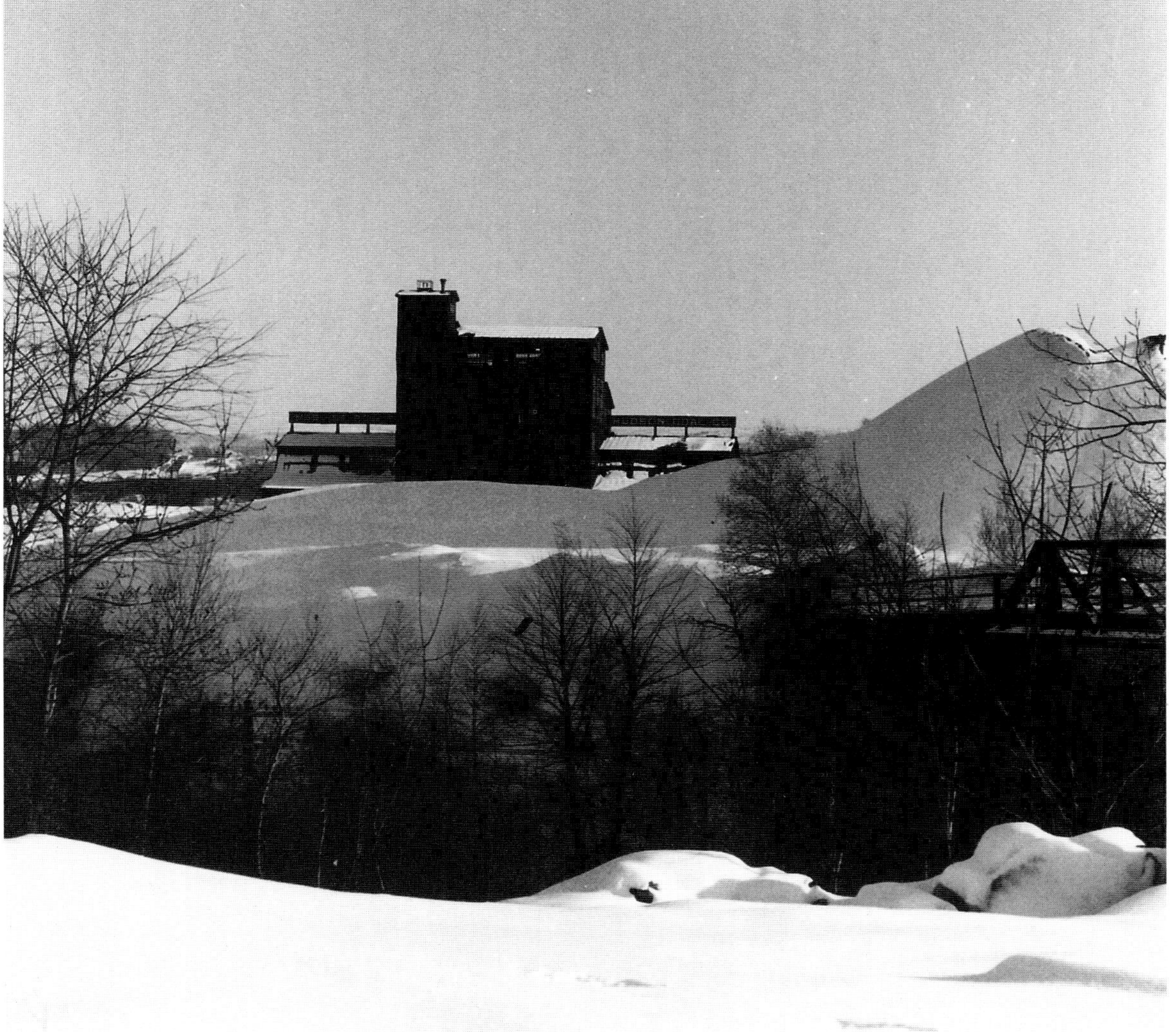

Marvin Colliery, Scranton; demolished in 1976.

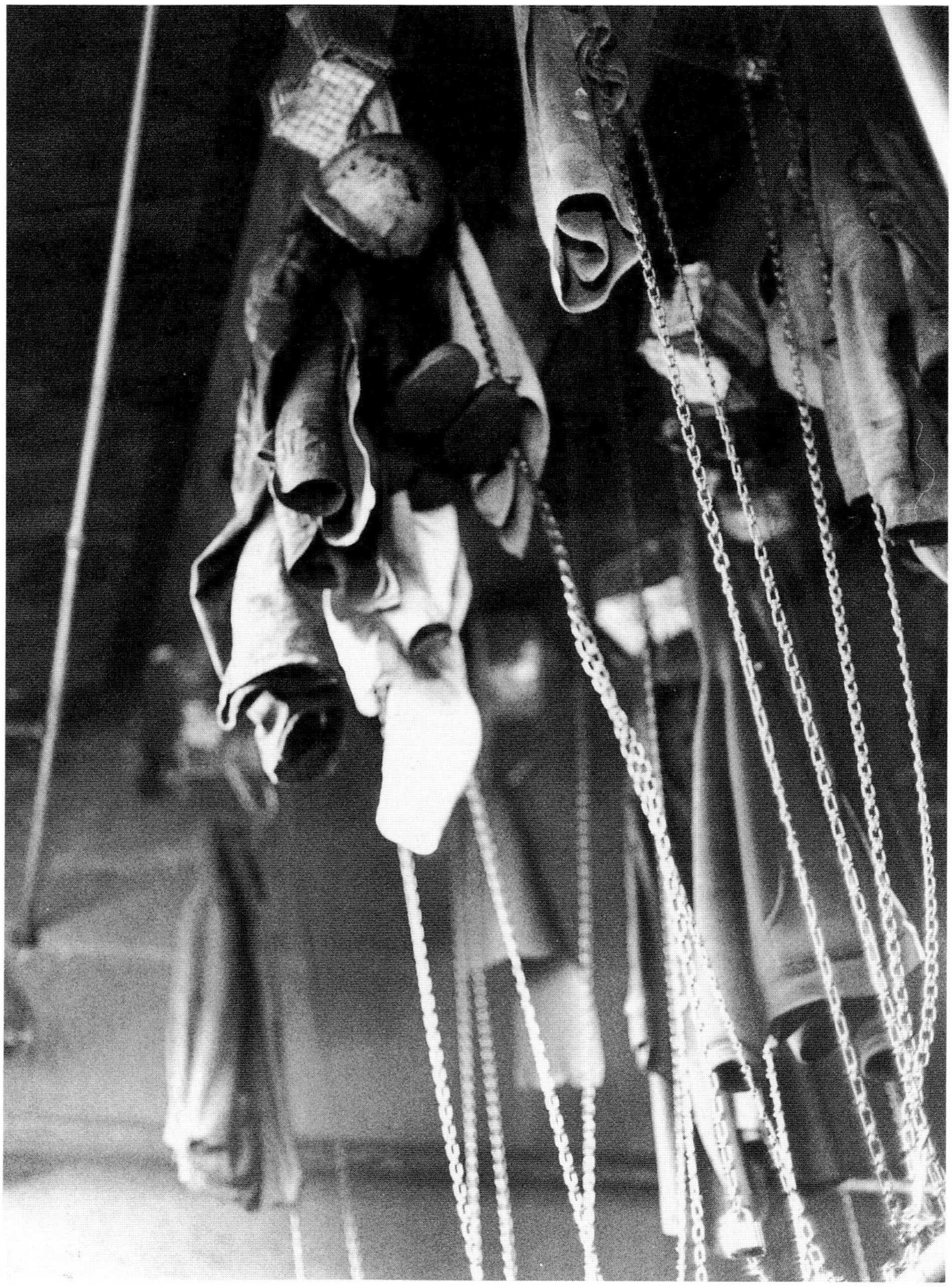

In an old wash house, boots still hang from where they were hoisted to dry.

"We called them shifting shanties, a place to shift from street clothes to work clothes. Shoes and clothes were hung up to dry."—Joe Sleyzak

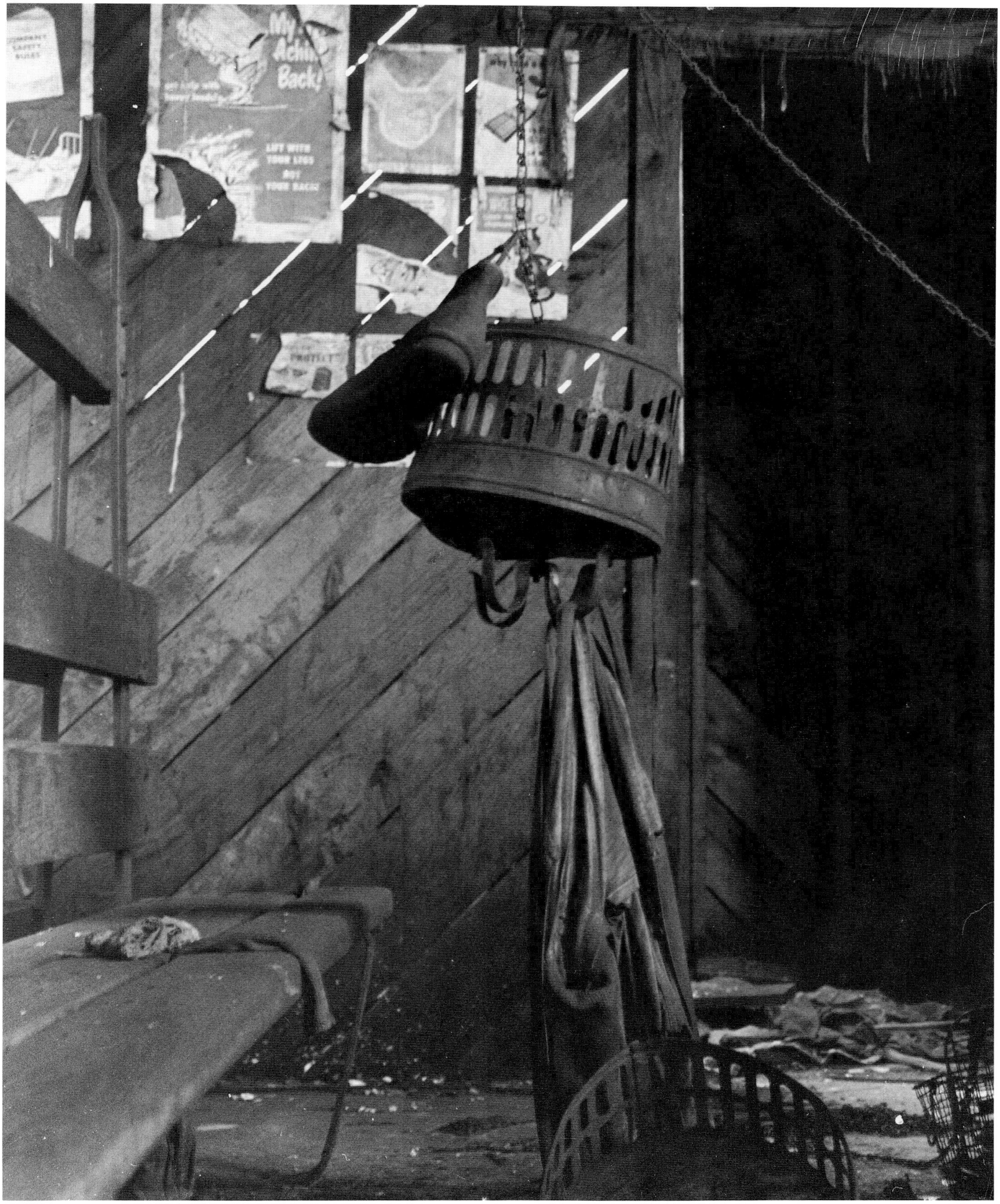

Interior of an abandoned wash house. The metal baskets contained the miners' work clothes.

"A miner worries about his work. It is a dread that is always with you. If Jake and his roof-bolt machine do not do their job properly, the mountain may cave in on the miners. Jake worried about his machine. I know of many miners who have been killed or injured in the mines. I would say that 75 or 80 per cent have been hurt in some way in the mines."—Joe Sleyzak

Records compiled in 1947 disclosed that during 100 years of recording fatal mine accidents, more than 120,000 men died violently in coal mines, averaging 100 every month for a century. The total does not include those who died of "natural" causes, such as black lung disease, in work notorious for its hazards to health as well as to life and limb. The "natural" death rate in 1947 among men aged 60 to 64 was eight times that of workers in any other industrial occupation.

. . . about 25% of all the men employed underground developed anthracosilicosis (miners' asthma) after a working period of more than twenty-five years. . . . . The prevalence of miners asthma among the entire group was found to be about 23%.—Pennsylvania Dept. of Labor and Industry, *Anthraco-silicosis* (miners' asthma): *A Preliminary Report of a Study Made in the Anthracite Region of Pennsylvania,* 1924

The 'black spit' and 'black lung' of the miners and the problems connected with them, are the most eloquent expression of intimate contact. . . . . The great enemies of the miner have always been water, foul air, and poisonous gases.
—George Rosen, *History of Miners' Diseases,* 1943

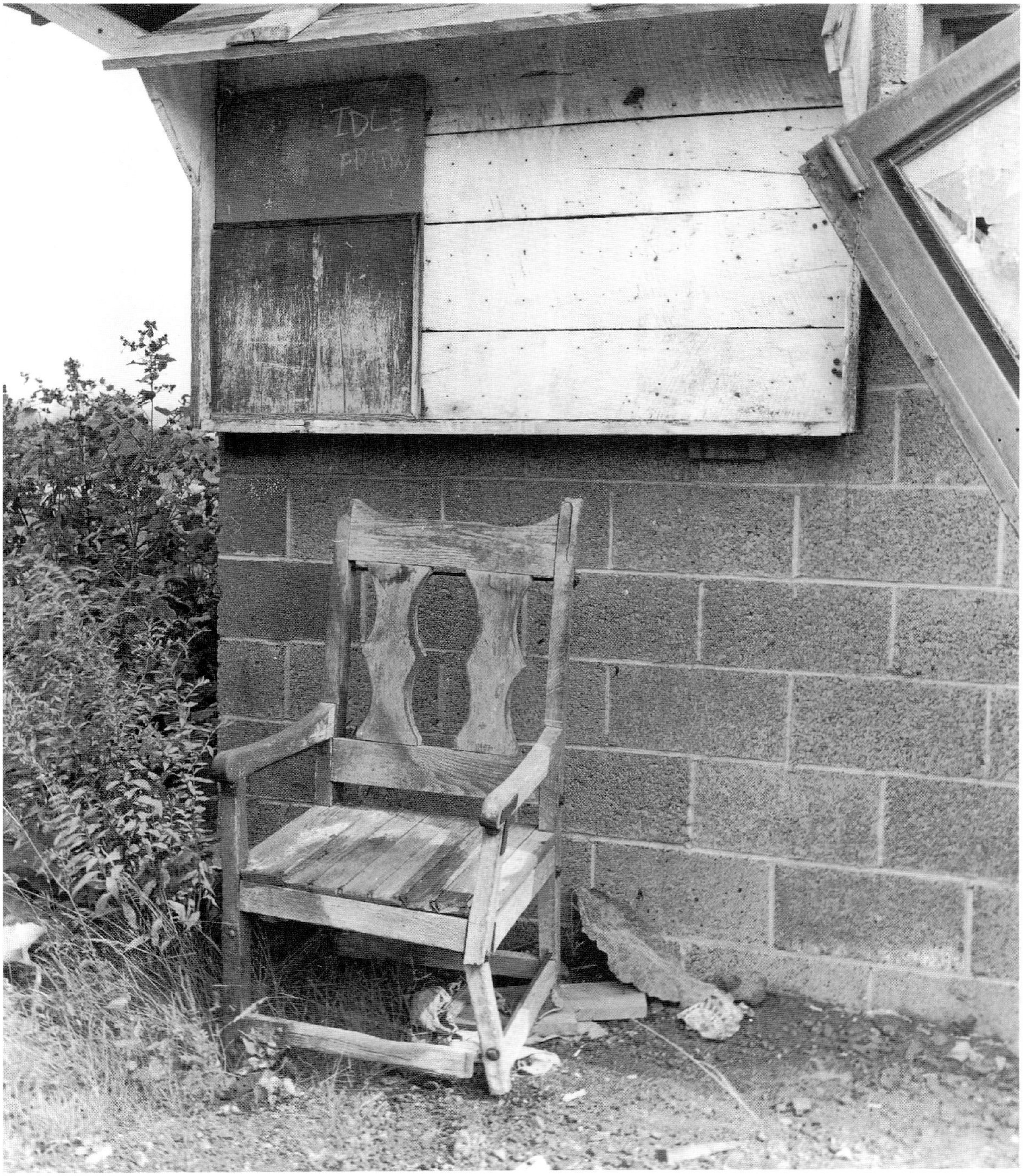
IDLE

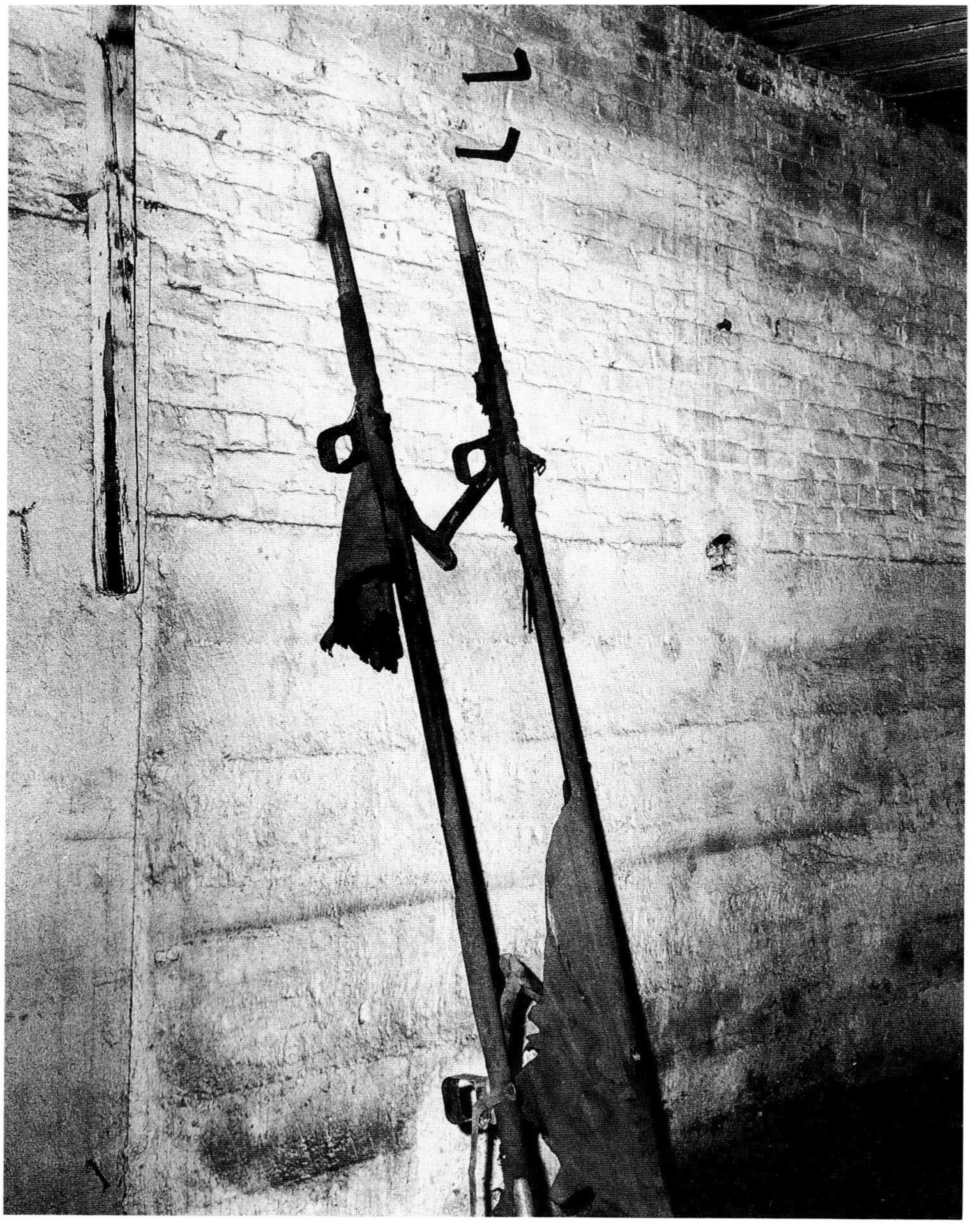

A rotting stretcher leaning against a wall where it was left when the colliery closed.

We passed a little shed, an office of some sort. A poster was pasted on the window, lettered, "Danger! Be Careful!" over a picture of a miner half cut in two by a machine.—Laura Gilfillan, *I Went to Pit College,* 1934

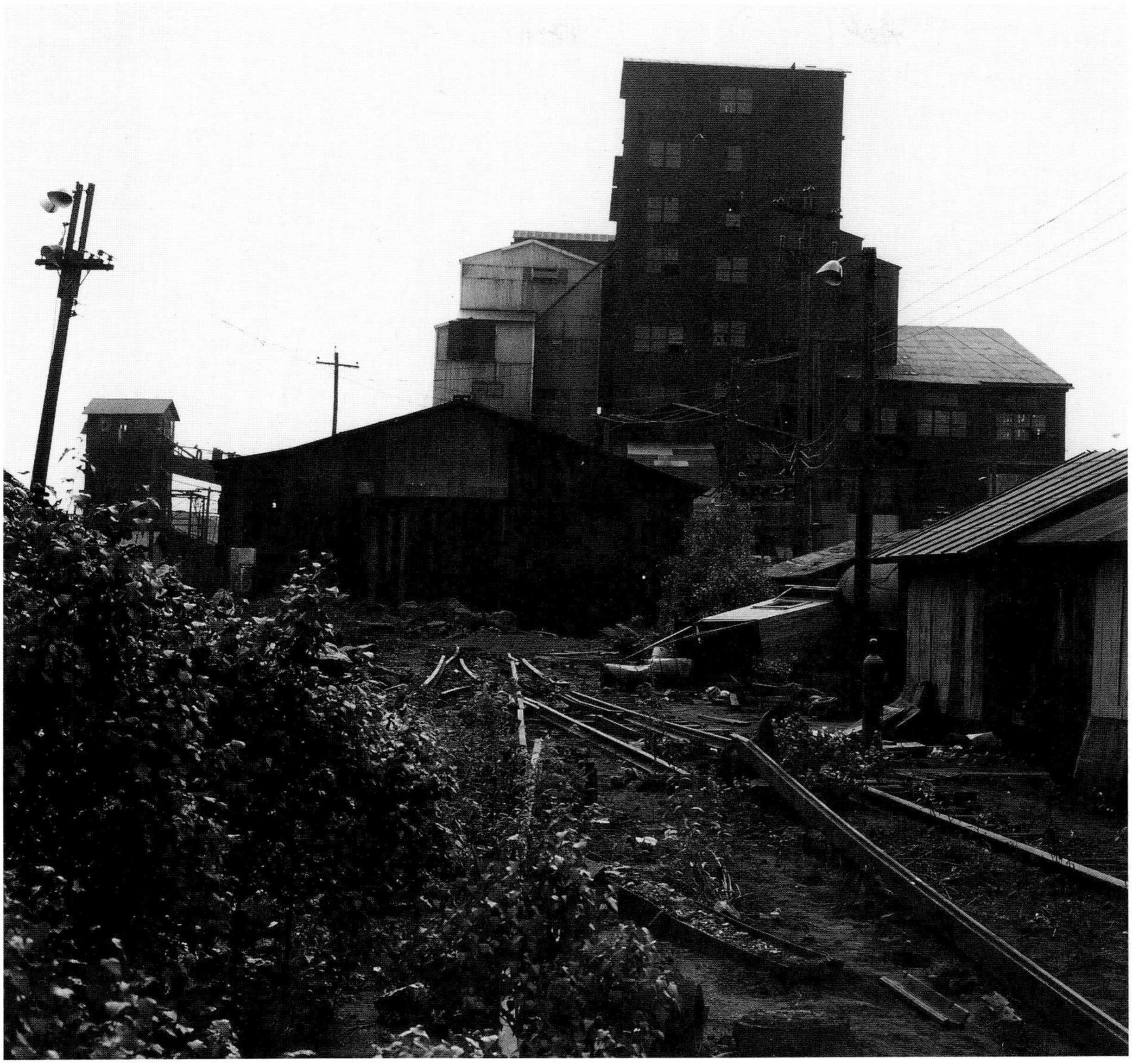

Jeddo #7 Breaker at Hazelton, 1973.

Hubbabub Hullabaloo
Two hundred cars will do.
Hoist away, every day
Clean the coal; make it pay.
In Freeland so frisky
They sell ice cream and whiskey.
So whack for old woodside,
It's my darling old place.
—Jack Johnson,
coal mining song, 1870s

"I started working in the mine at 14. Then miners used hand machines to drill holes. Then they went to jackhammers. Hand machines kept dust down to a minimum, but jackhammers, you really clouded the black dust there . . . that's what got to most of the men. You got the hole drilled in a shorter time and you got your day's pay in less time . . . but it wasn't worth it in the long run.

When I first started to work in these things [mines], they used to crush the coal dry and you never seen such dust; what a filthy job that was. It was a livelihood at one time, but it's an ugly picture today."—Joe Sleyzak

My father worked in the mines . . . was lucky if he was paid as much as one dollar per day for ten hours work. My mother was the daughter of a coal miner and in this environment no choice was presented to me other than timber or the coal mines. As a trapper boy, I received 25 cents per day for a ten hour shift. After being on the job for a short time, I was promoted to attend two trap doors and tend the fire in the ventilating furnace. For this work I was paid 75 cents per day. I imagined myself well on the road to wealth.—Fred Mooney, *Struggle in the Coal Fields,* 1967

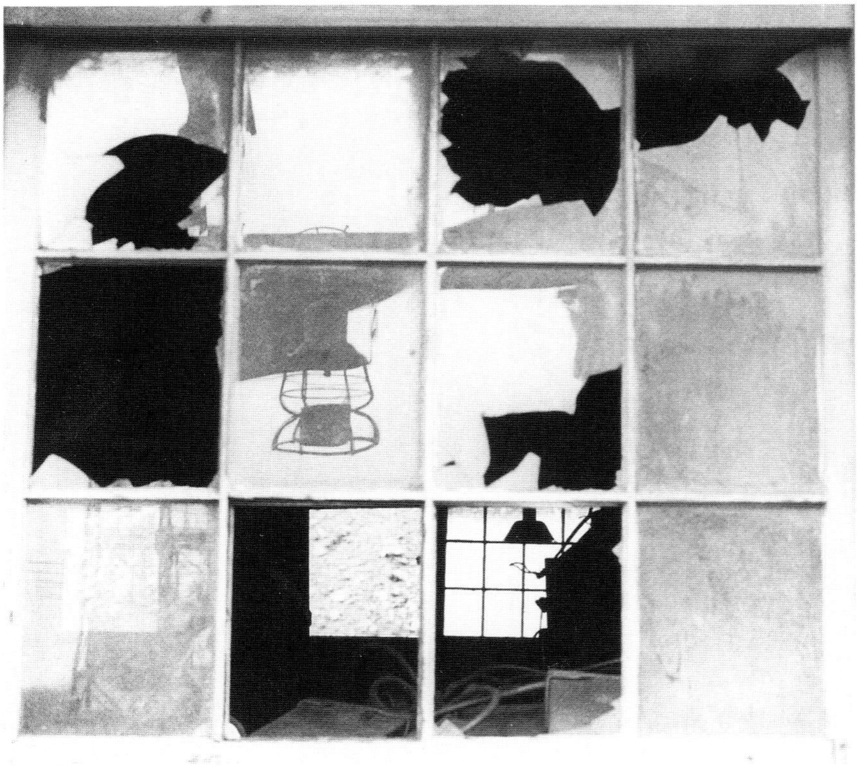

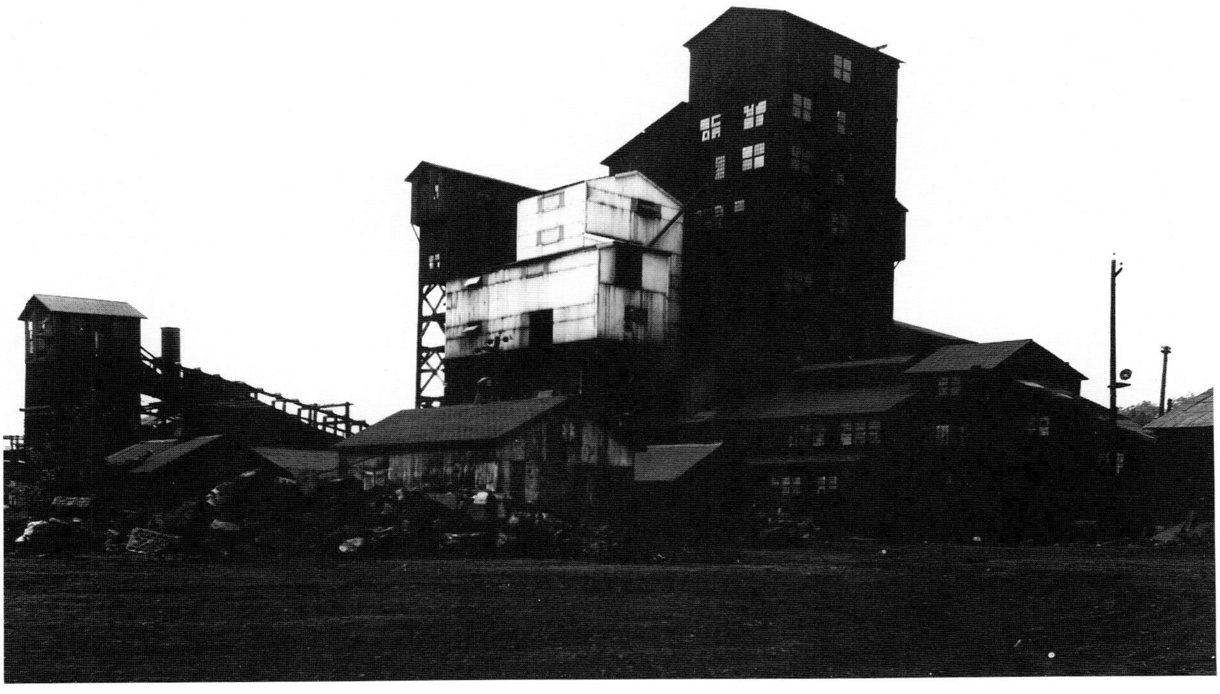

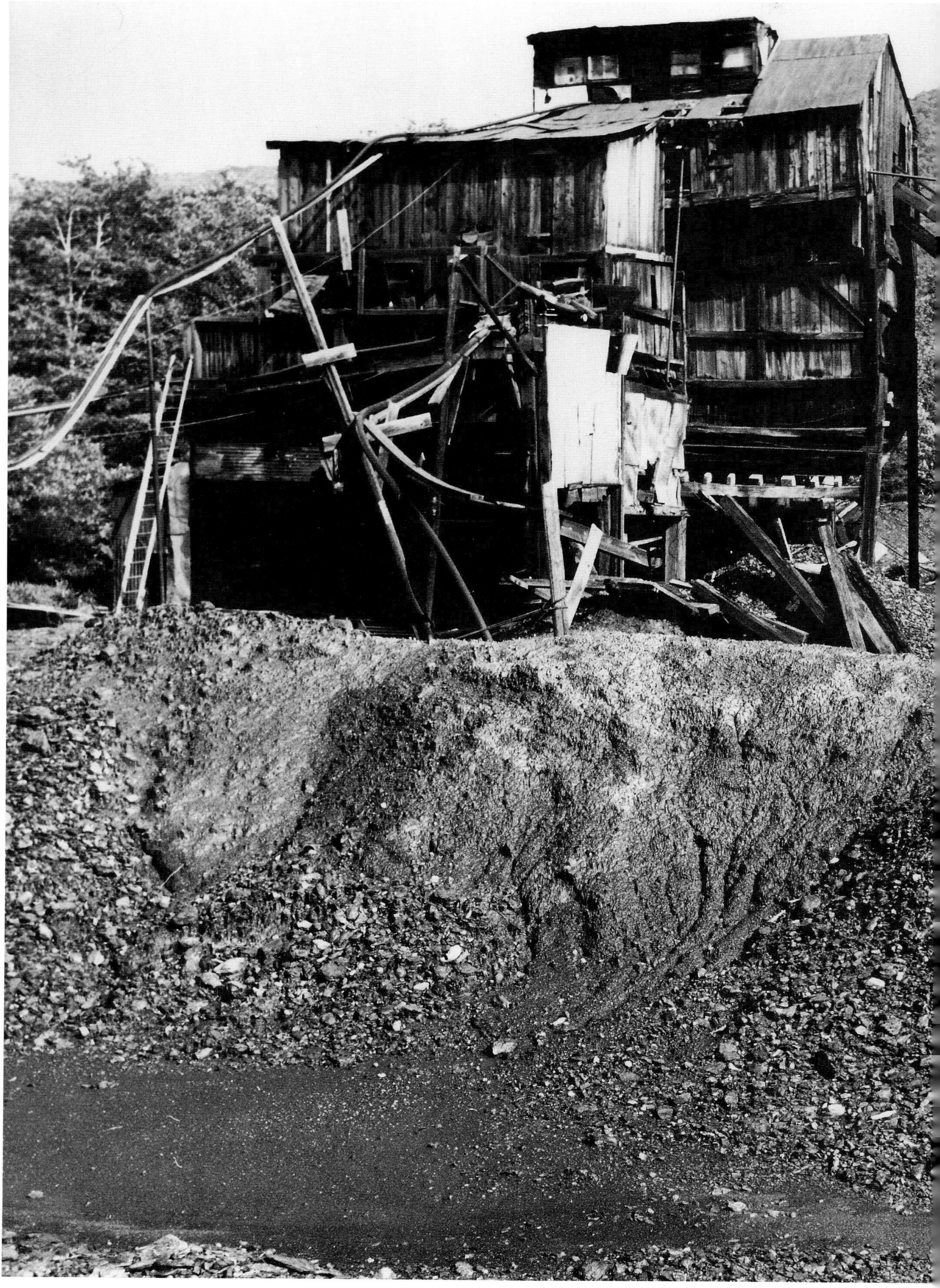

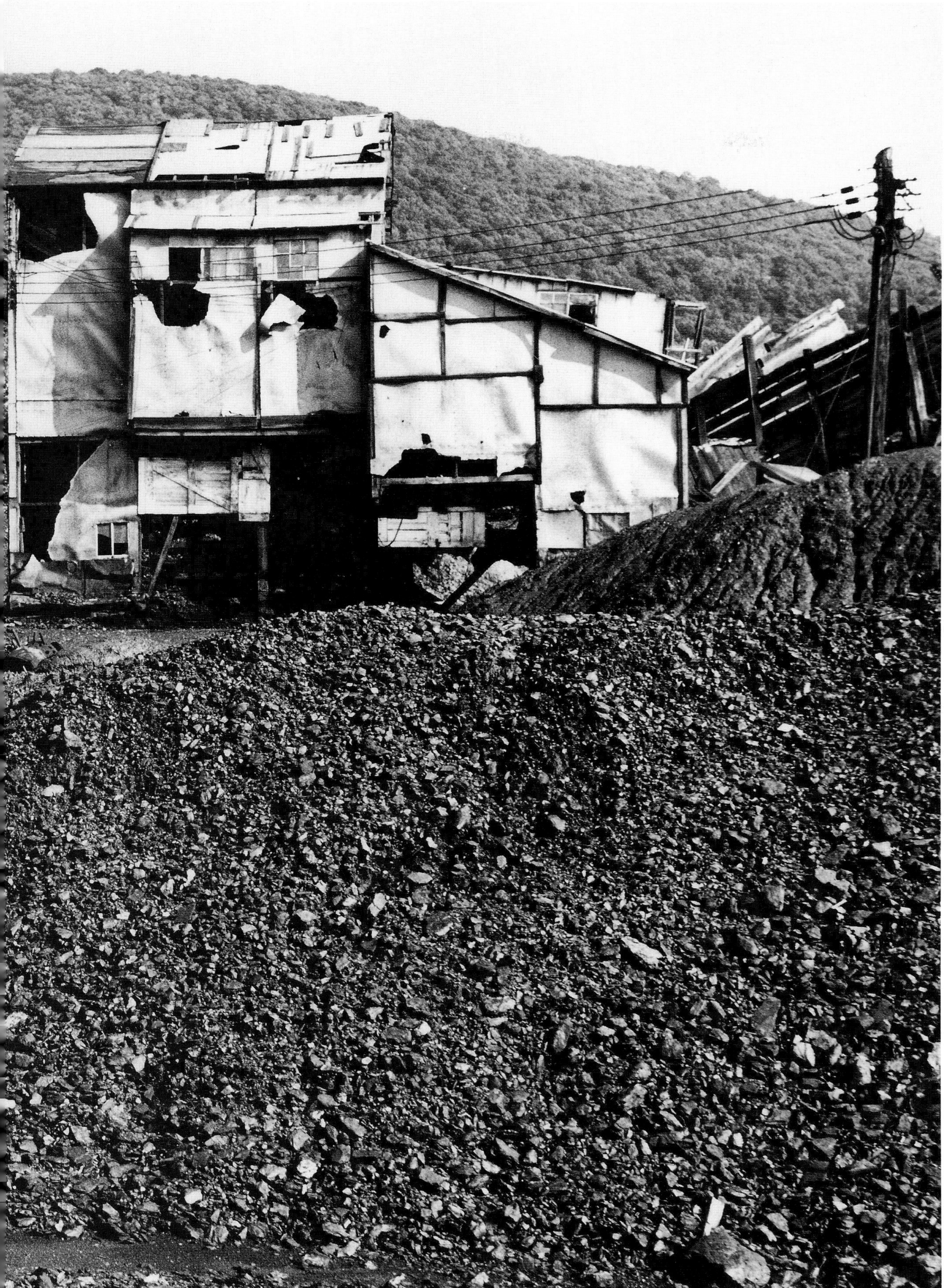

I'm a hard working man, you can see by my hands,
    Although I am friendly and free.
A dollar a day is very small pay
    For a man with a large family.
I work in the mines where the sun never shines
    Nor daylight does ever appear;
With me lamp blazing red on the top of my head,
    And in danger I never know fear.
Just think of the poor man who works in the mines
    With the mules and the rats underground;
Where the smoke is so thick you can cut it with a stick,
    And can weigh it on scales by the pound.

—Patrick J. "Giant" O'Neill, "The Hard Working Miner,"
Songs and Ballads of the Anthracite Miner

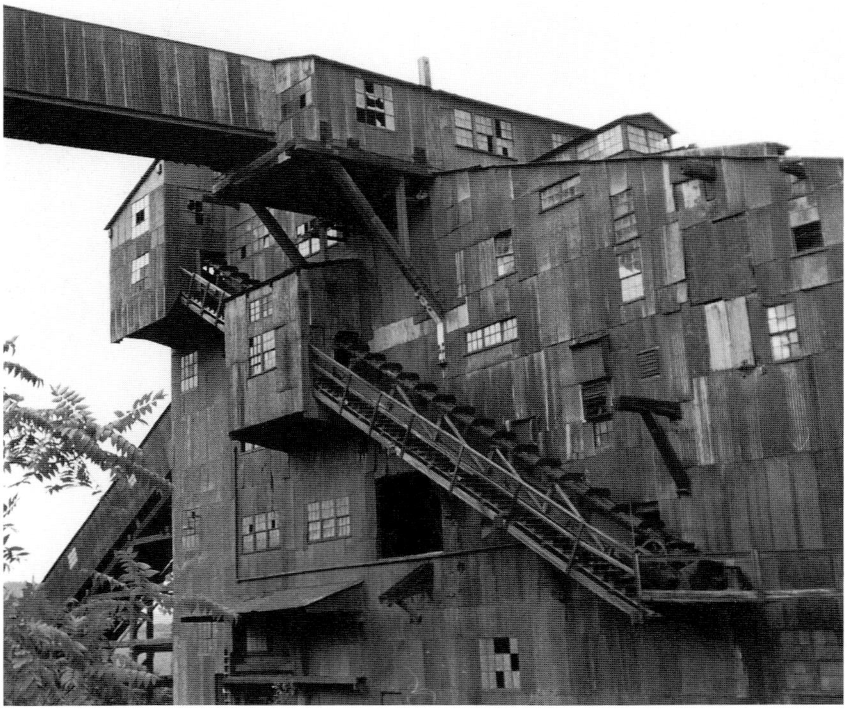

Overleaf: Both the colliery and the environment are in ruins.

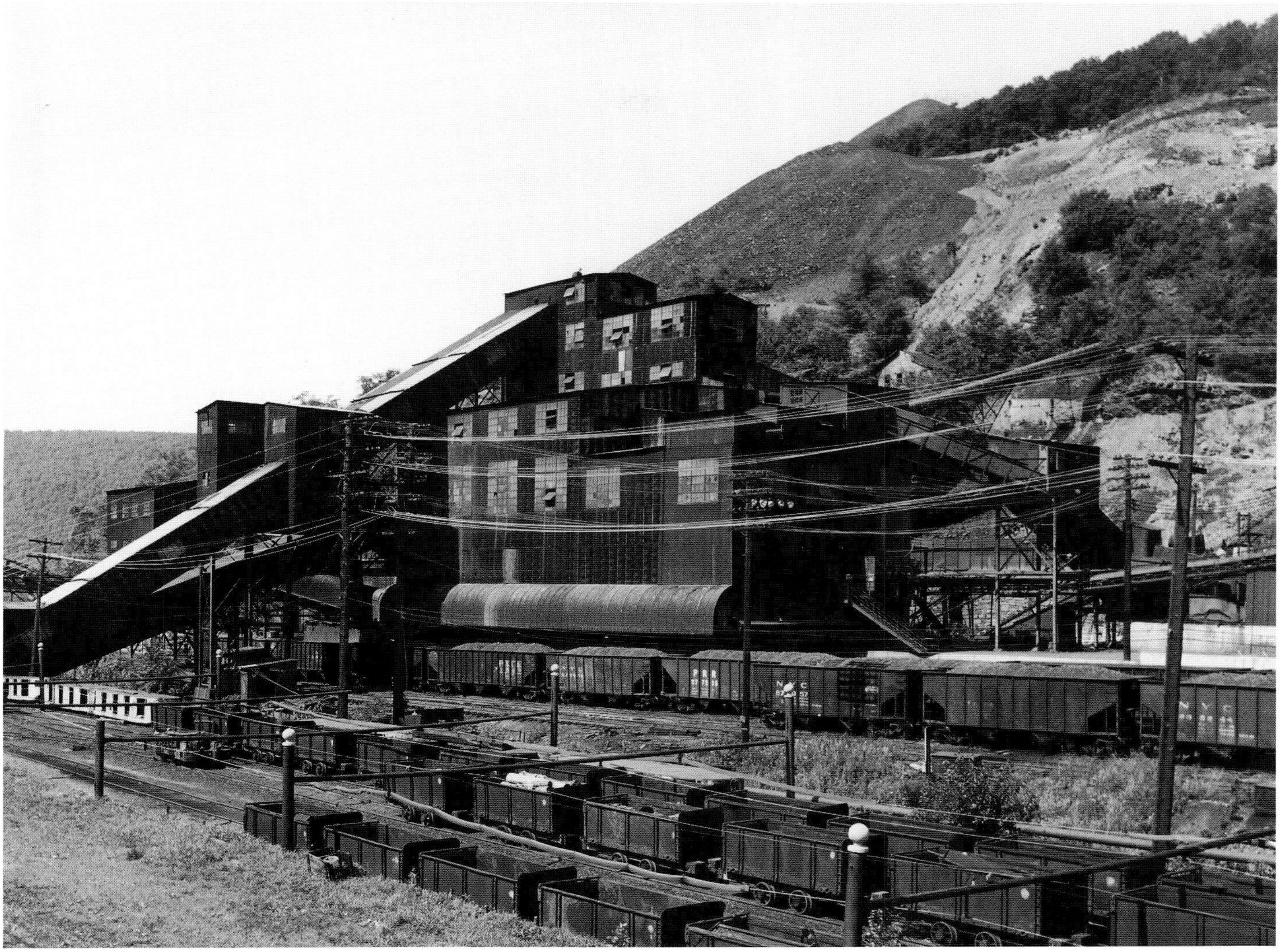

Glen Burn Colliery with the breaker below the highest and longest-burning culm bank, Shamokin, 1973.

Our troubles are o'er Mrs. Murphy,
    For the Dutchman next door tells me straight
That the mines will start full-time on Monday.
    That's what he tells me 'tanny rate.
Sure, the boss, he says, told him this morning
    As he was 'bout entering the mine,
That the coal is quite scarce roun' 'bout New York
    So the rumors is work full time.
—Con Corban, 1880s

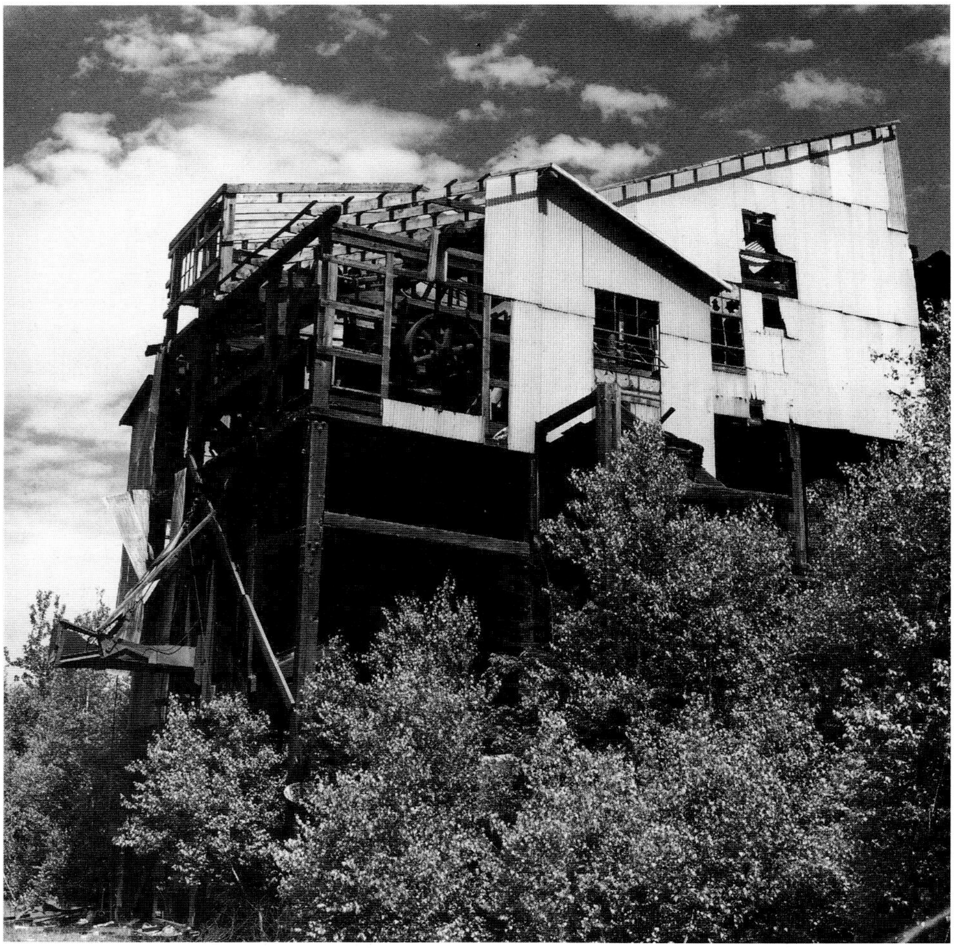

When I raised my family in Chattaroy Holler
The coal train run right by my door.
Now it's sittin' in the yard all rusty and empty
And the N&W train don't stop here no more.
—Nimrod Workman, "The N&W," *Voices From the Mountains*

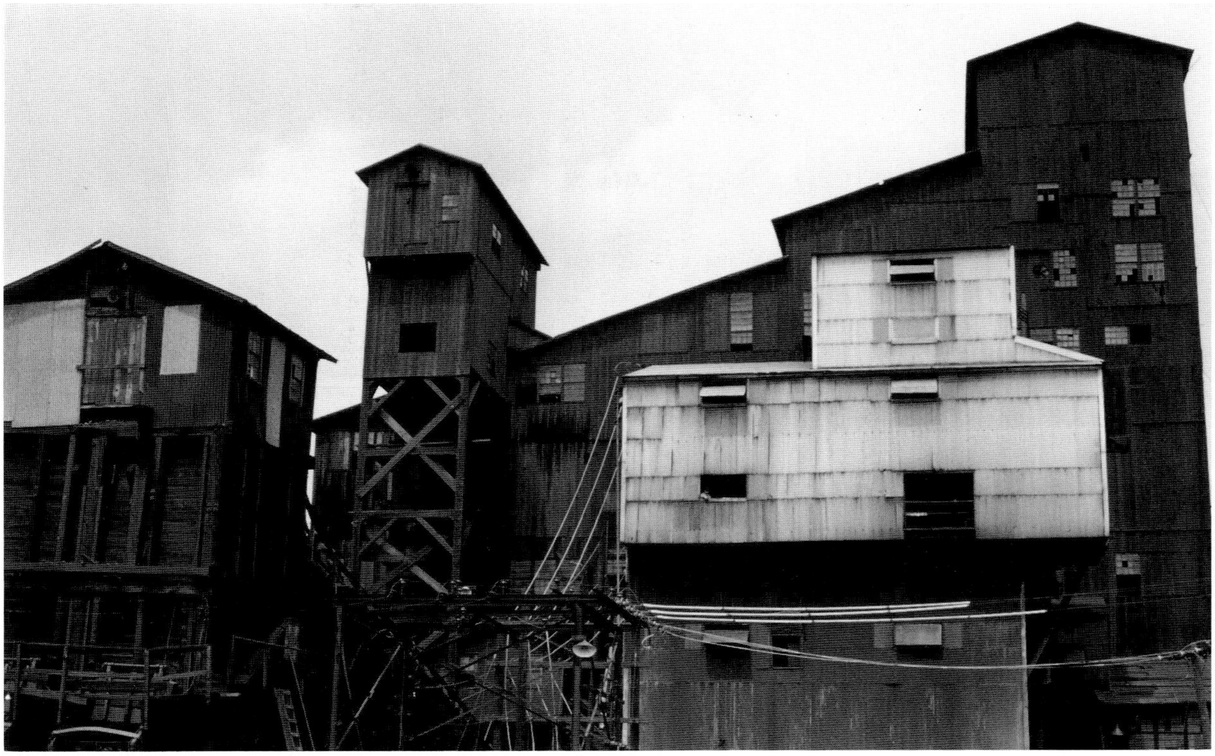

Jeddo #7 Colliery, Hazelton, 1973.

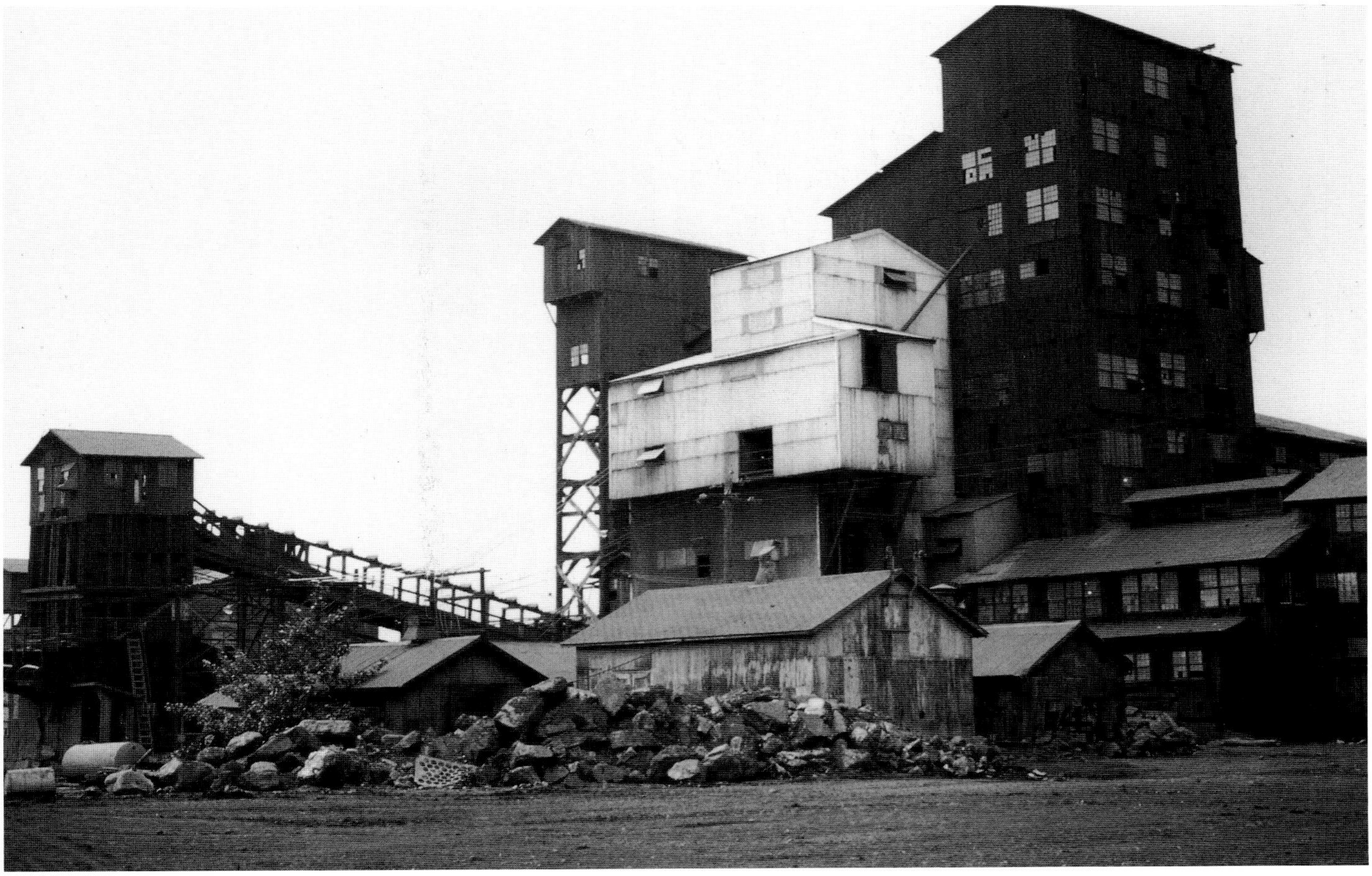

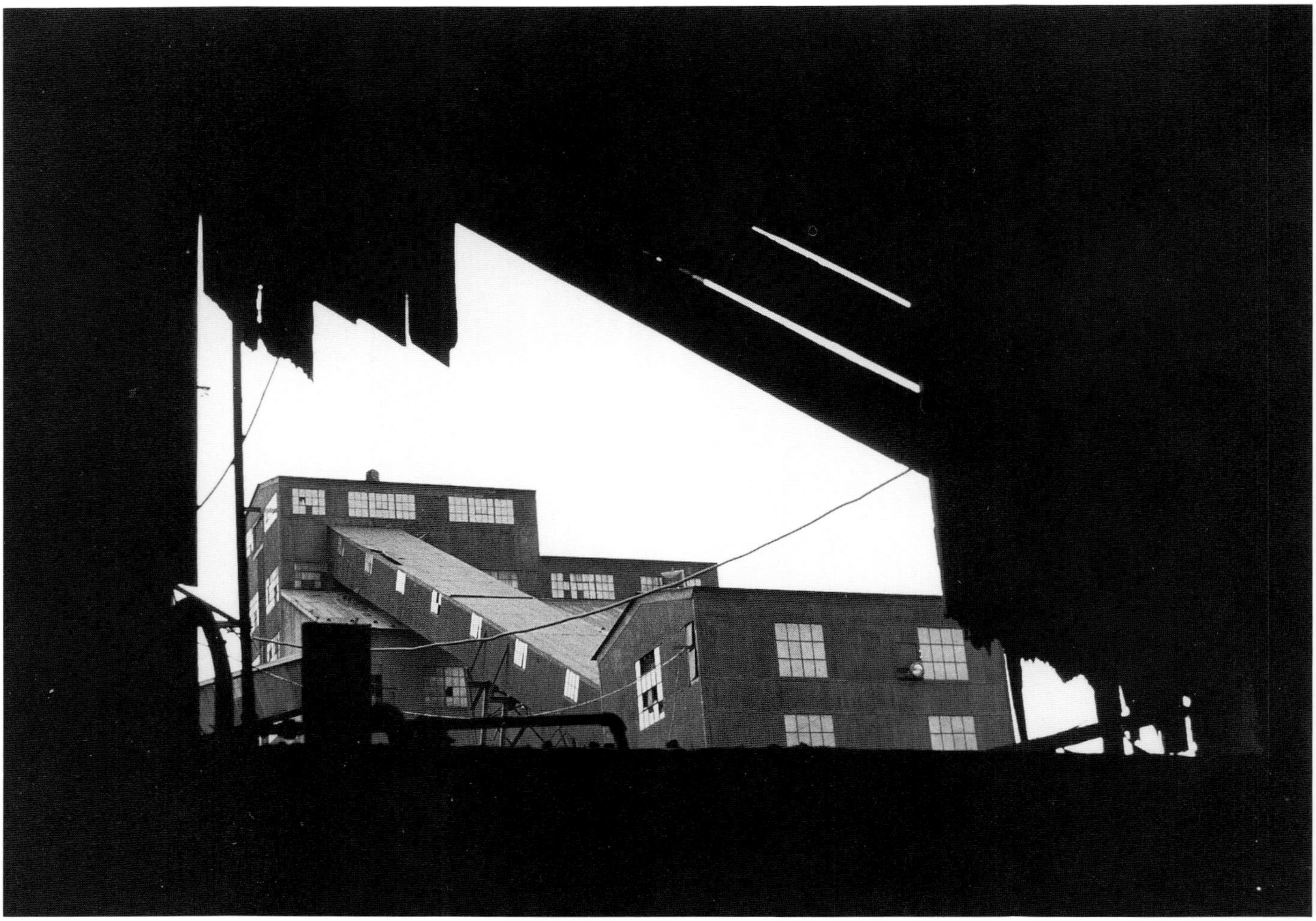

The miner was paid for lump coal only, all that passed through the screen being deducted in weighing. The screenings were sold by the company in two grades known as 'nut' and 'pea' at reduced rates, and the miner thought that as he had mined and loaded the coal, he was entitled to a proportionate share of the sales.—Andrew Roy, *A History of the Coal Miners of the U.S. . . . to the Anthracite Strike of 1902*

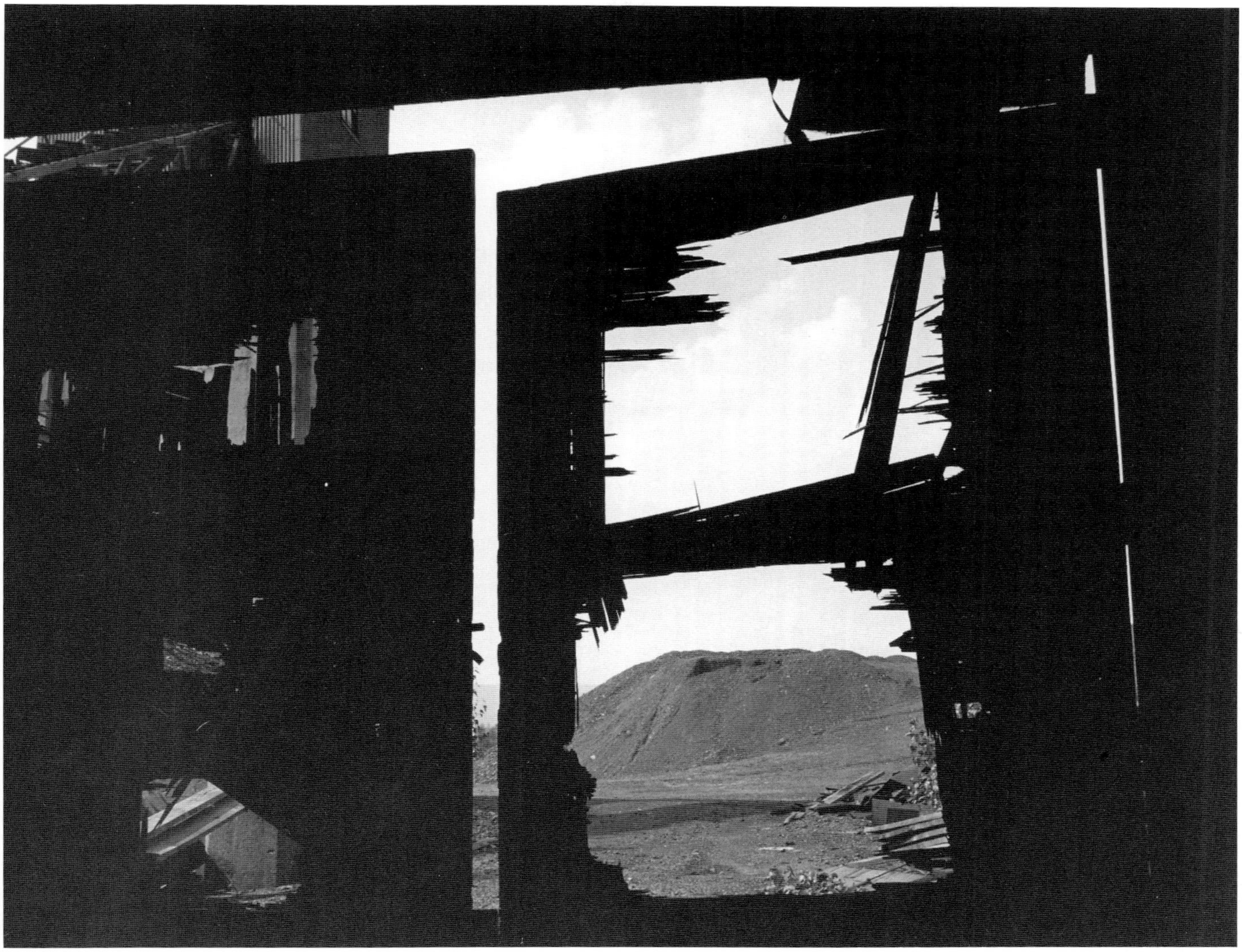

My sweetheart's a mule in the mines,
I drive her without any lines,
On the bumpers I sit
And tobacco I spit
All over my sweetheart's behind.
—G. G. Korson, *Minstrels of the Mine Patch,* 1927

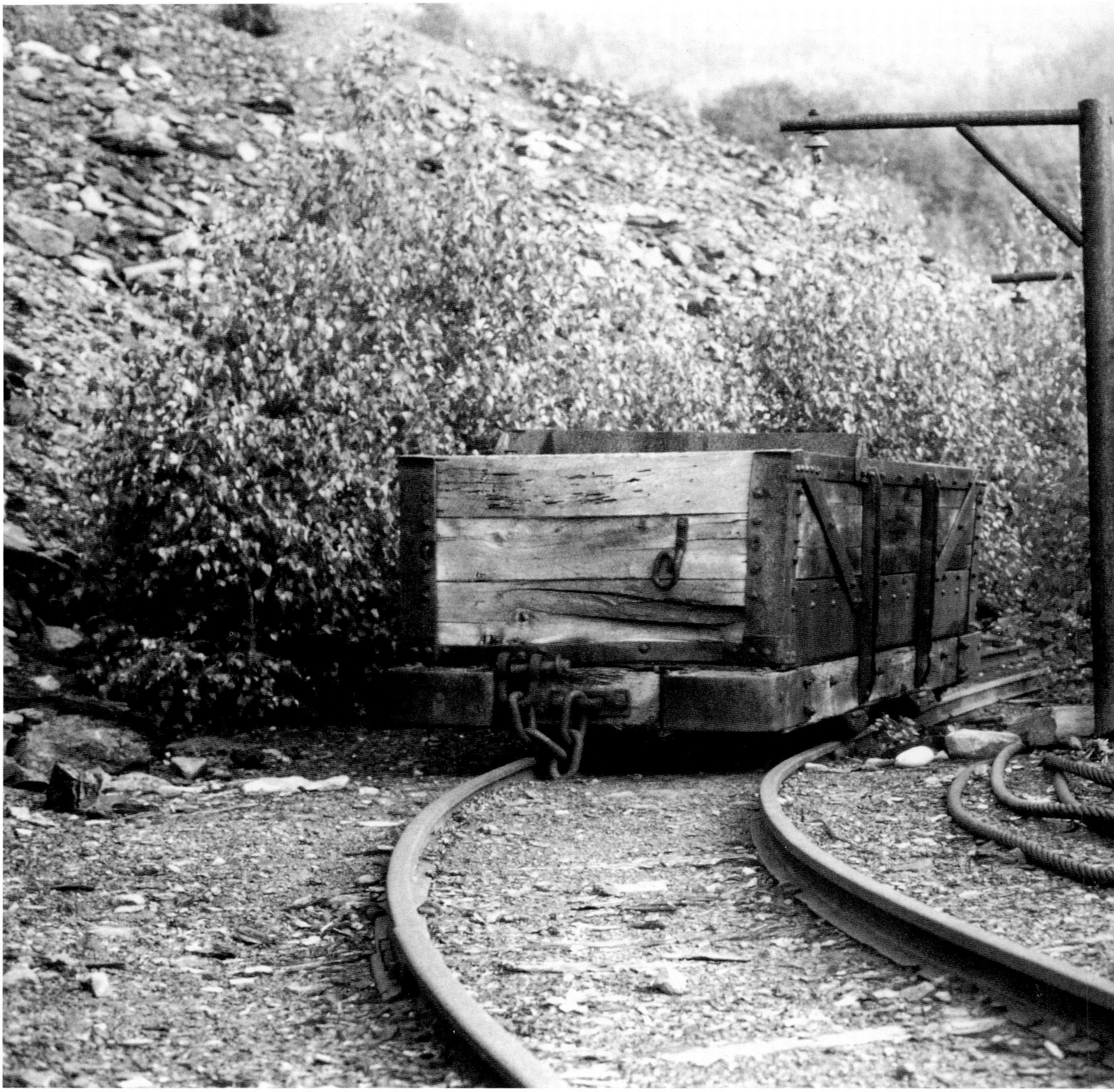

A lone coal car stands near a closed mine.

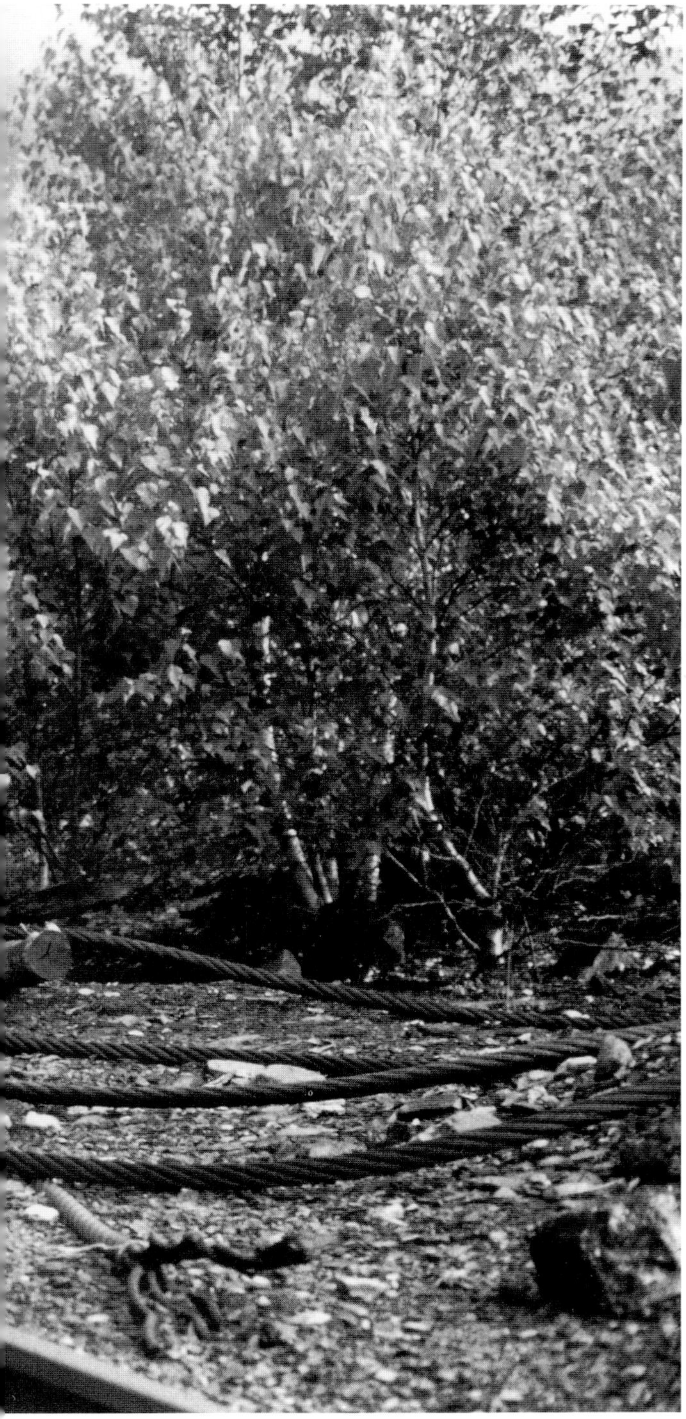

The cars in which the miners sent out their coal were not of uniform capacity nor properly branded. New cars sent down into the mine by the operator would be larger than the old ones, while the miner would still be receiving payment according to the old measures. What constituted a ton of coal was left to the discretion of the operator. An 1864 petition for correction was postponed.

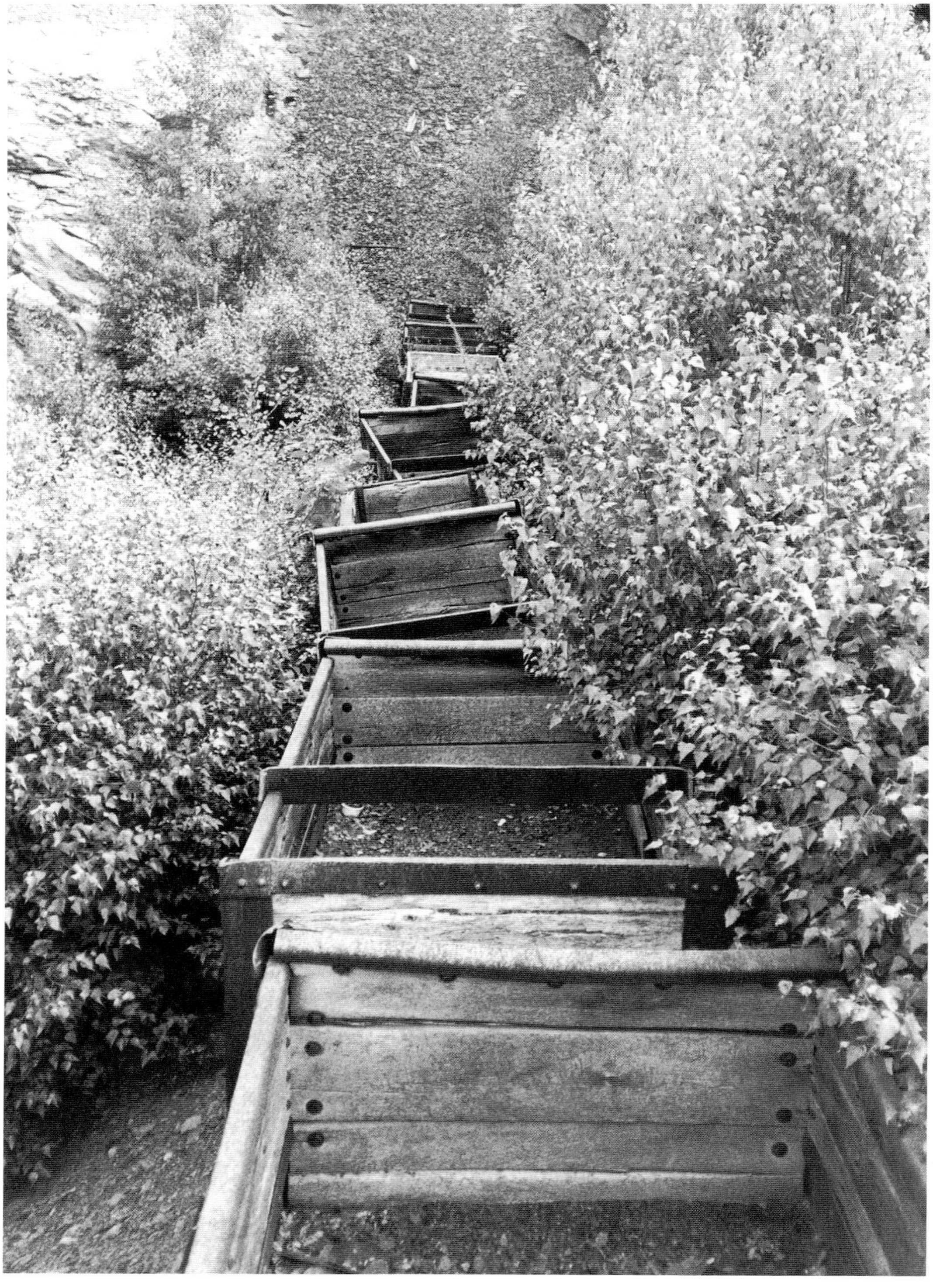

A line of coal cars heading into a coal mine. The lead cars are buried in an avalanche of coal used to seal the mine for safety reasons.

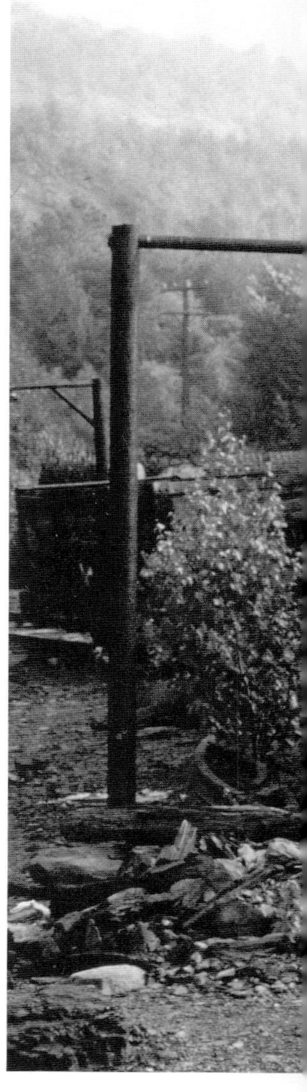

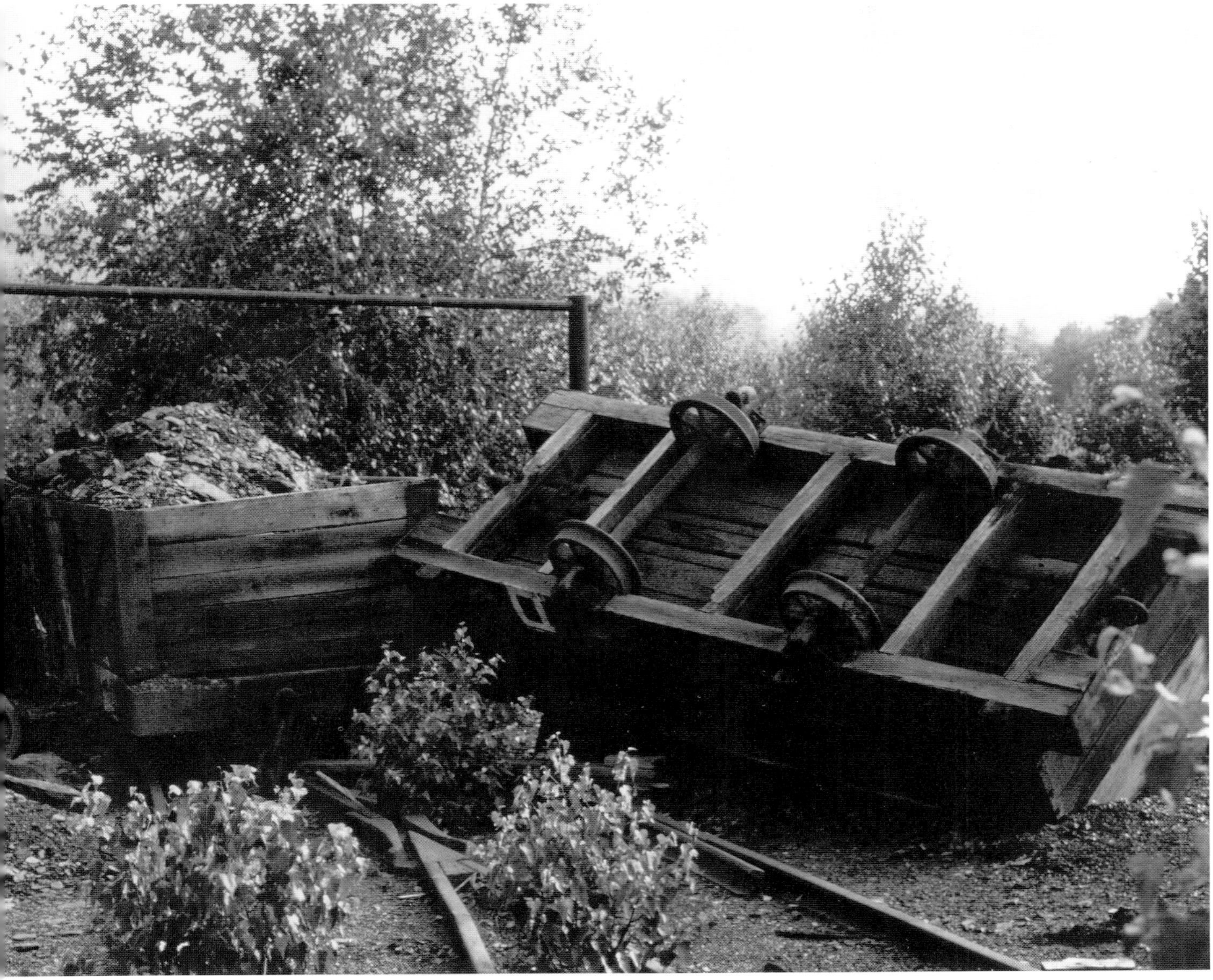

A coal car was derailed when the colliery closed, leaving a few cars standing, still full of coal as it came from the mine.

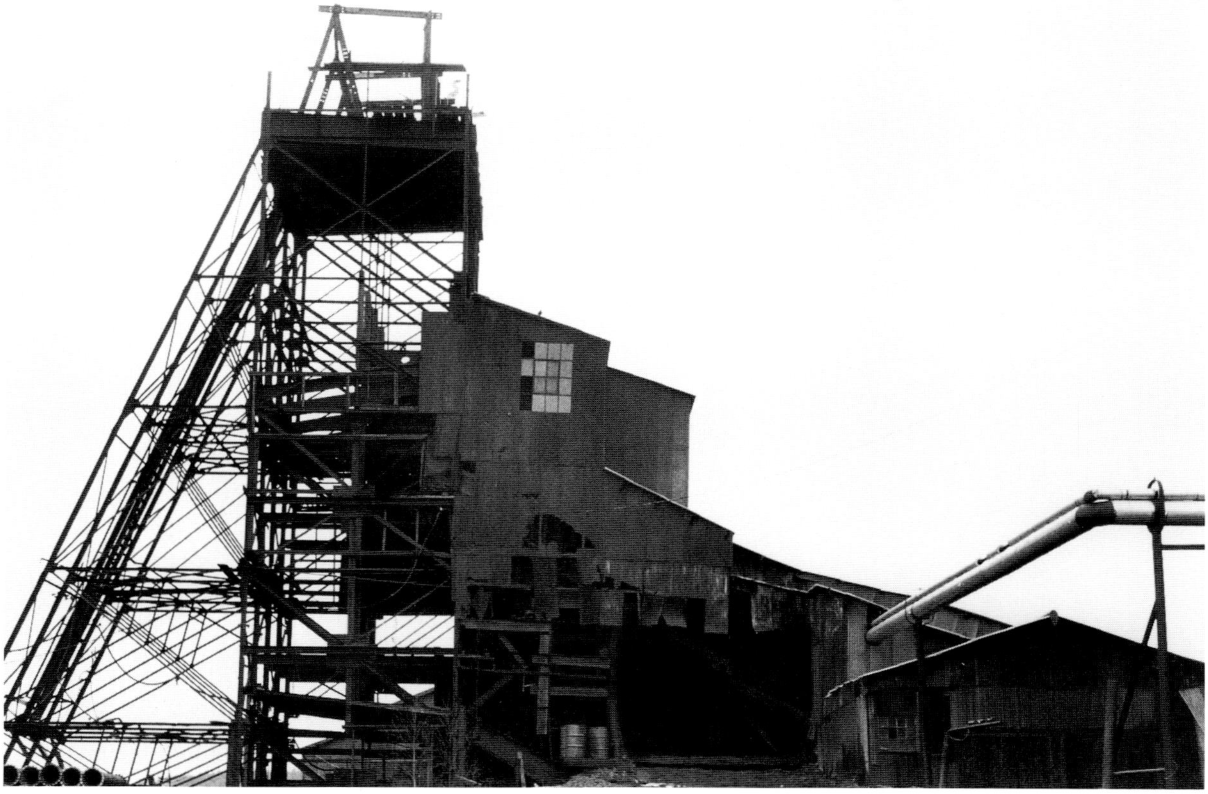

A tipple over a mine shaft.

I am a jovial collier lad, and blithe as blithe can be,
For let the times be good or bad, they're all the same to me;
'Tis little of the world I know and care less for its ways,
For where the Dog Star never glows I wear away my days.
Down in a coal mine, underneath the ground,
Where a gleam of sunshine never can be found;
Digging dusky diamonds all season round,
Down in a coal mine underneath the ground.
—J. B. Geoghegan, "Down in a Coal Mine," 1872

Our wages, John, grow beautifully less,
And if they keep on growing thus, I guess,
We'll have to put on magnifying specs
To see the little figures on our checks.
It's nothing strange to find on seeing the docket,
We've worked a month and still are out of pocket.
It makes a man feel dirty cheap, you bet,
To work a month and then come out in debt.
—Anthony Bimba, *The Molly Maguires*

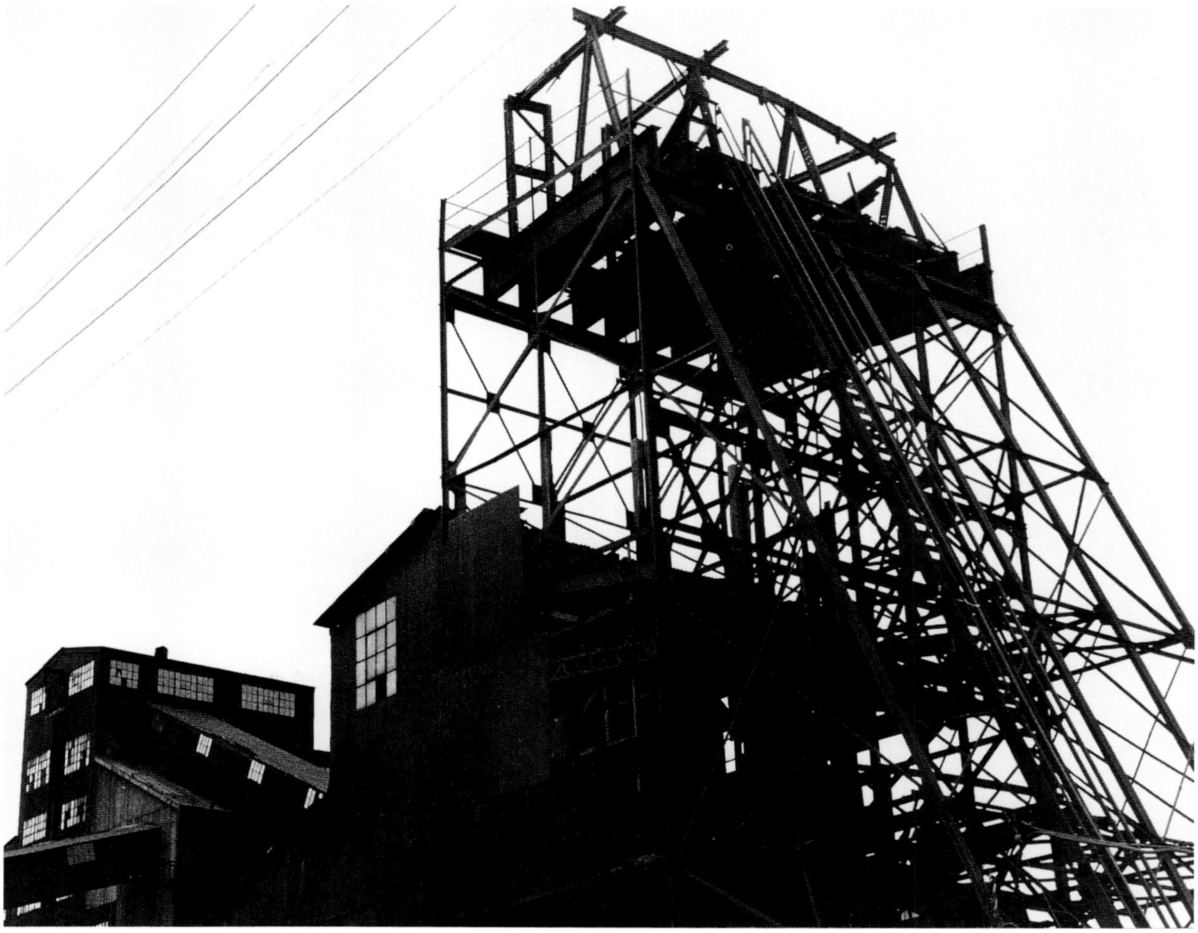

A tipple over a mine shaft.

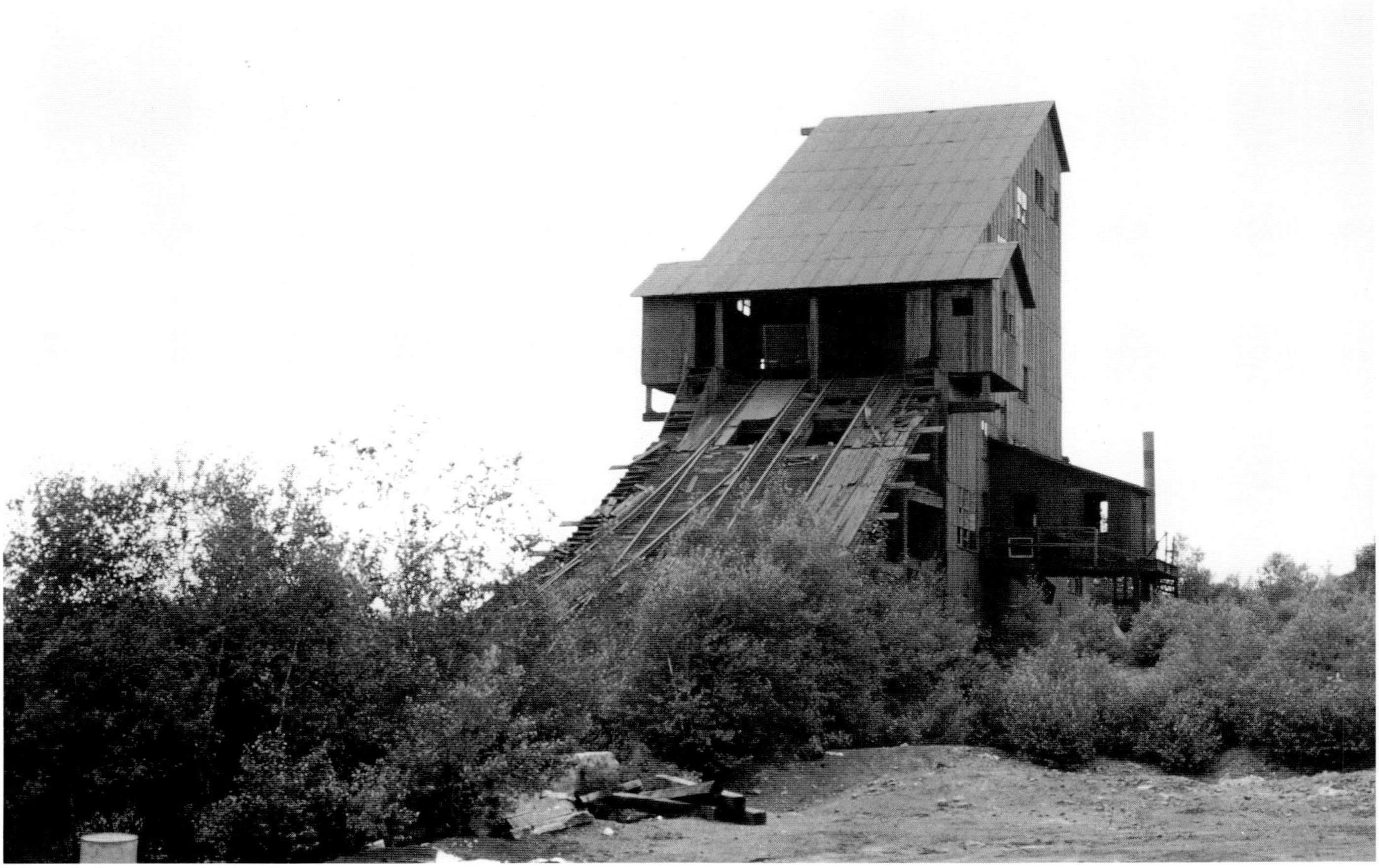

A covered tipple, Eberville, 1973.

The company stores charged double for everything and you had to pay in the scrip the company paid out instead of money. The scrip could be spent nowhere but at company stores. "You never make enough to pay the company all you owe, so they charge you all you make."

In 1843, petitions against the payment of wages in store orders were received by the Legislature, but nothing was done for the next twenty years. The strike in 1849 was against being paid in store orders.

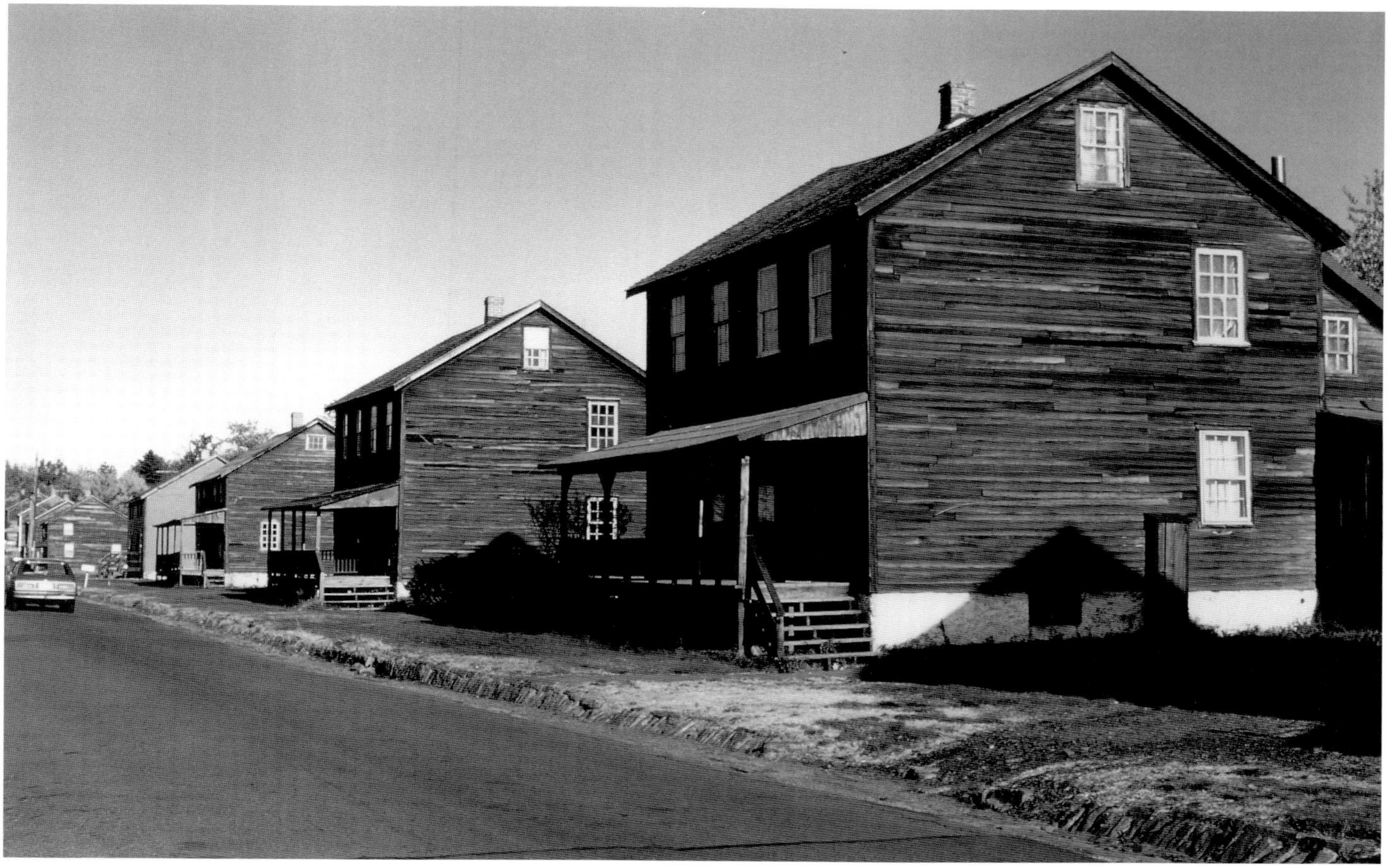

Eckley Mine Patch.

COAL MINE PATCH

Life was stark, bleak and semi-primitive. No houses were ever built more uninviting and uncomfortable than the first houses the companies built for the miners. In long rows, all alike, made of plank and batten, they kept out neither rain, heat, or cold. They had no plumbing. Water was obtained from a well or from community hydrants which froze in the winter. Lighting was by tallow candles or kerosene lamps. The streets were dark. Homes in the older patch towns were indelibly colored a dozen shades of gray by grimy coal dust, smoke, and soot.

In 1922, a mine patch house rented for $11.75 a month.

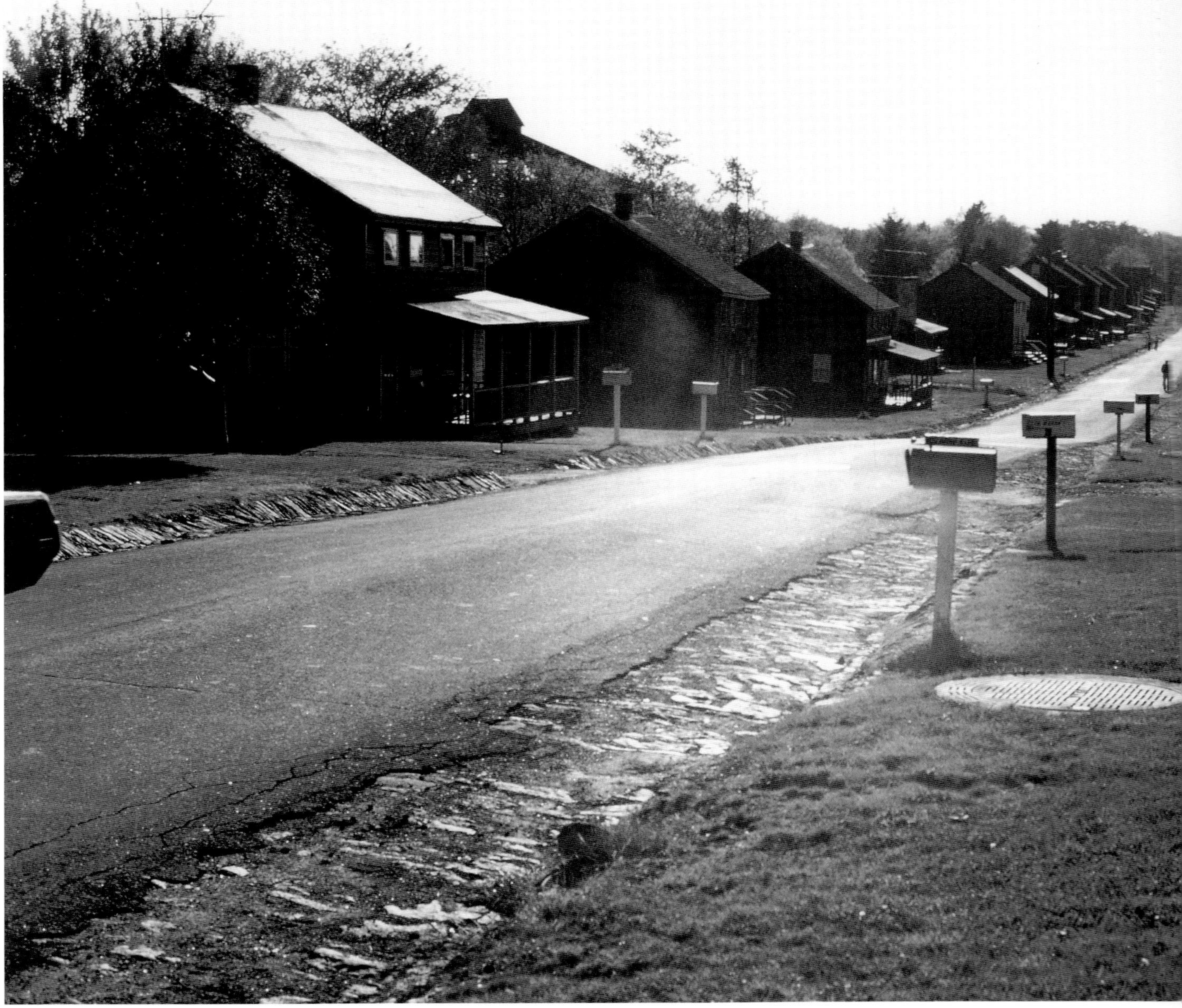

Eckley Mine Patch.

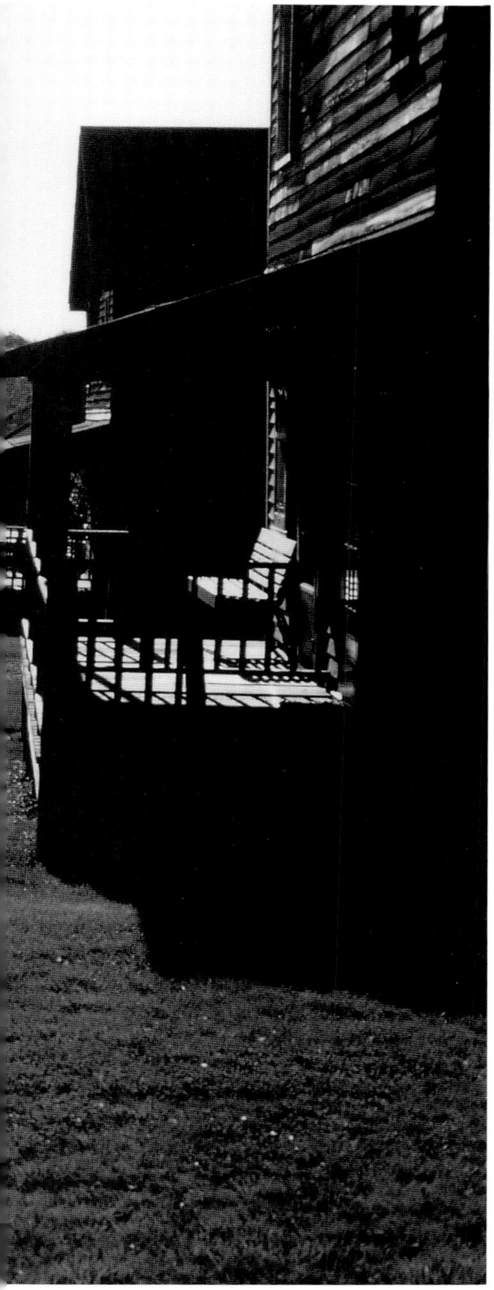

I was born in a coal patch, stars at my head,
And mountains as far as I could see.
Got to work every day and I made a big pay,
There was honey in the horn for me.
And the coal dust blowed, like a dark summer snow
But I don't hear the roar of the tipple any more.
—Billy Ed Wheeler, "Ain't Goin' Home Soon,"
Voices From the Mountains

Bitter was the miner's life.
In his sayings and songs he cursed it.
There's one thing I'll tell you and don't you forget
That your back with the droppers is all wringing wet,
Your clothes they are soaking, your shoes are wet through,
Oh! never be a miner whatever you do.
—Anthony Bimba

When I was a boy in Scranton,
The coal train run by my door.
Now it's sittin' in the steam museum, no more smoking.
The Delaware & Lackawanna train don't stop here no more.

. . . .

Deep in the bowels of the earth we sing.
We dig with a will, our picks we swing.
Shiny black coal is what we bring,
Dig, men, dig; hear our picks ring
With the echo made by rocks that cling
To the ebony coal we crown King.
So sing and swing, swing men, swing.
—*The Story of Anthracite,* A Marionette Play, WPA

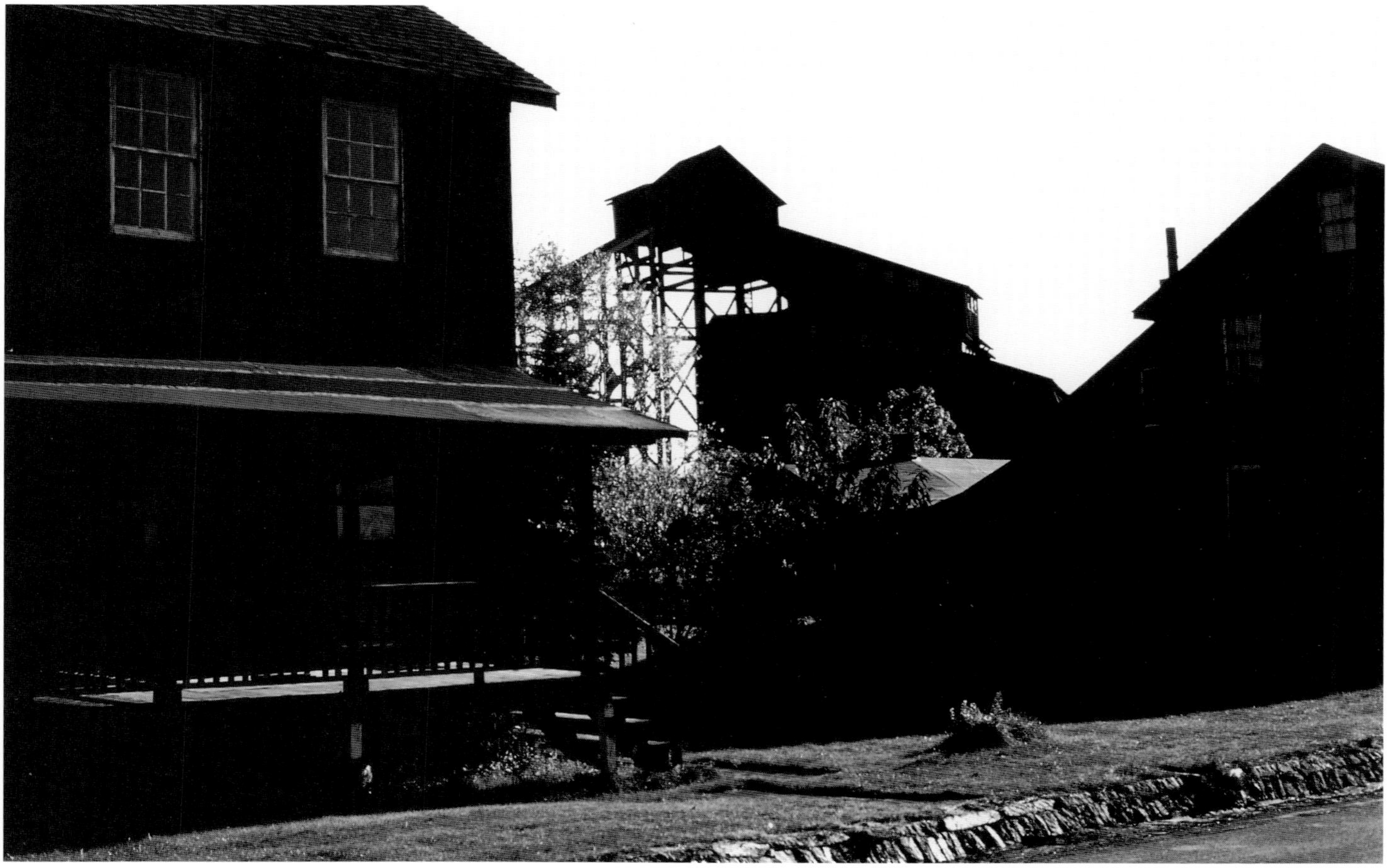

The breaker and the coal mine entrance was in the back yard of the houses in the Eckley Mine Patch.

A company patch, in the shadow of the slate dumps, the older patches climb hilltops in rows of identical houses with fenced back-yards and outdoor toilets. In 1920, with the exception of Hazelton, Freeland and possibly McAdoo, most of the mine workers in the Middle Field live in company towns and patches, or settlements near collieries. In some places, mining operations have crowded the towns very closely and there is little available space for dwellings. The country is characterized by small towns huddled together in the valleys or on mountains, surrounded on all sides by mammoth culm banks and coal breakers.—National Industrial Conference Board, *Cost of Living Among Wage Earners: Anthracite Region of Pennsylvania*, 1922

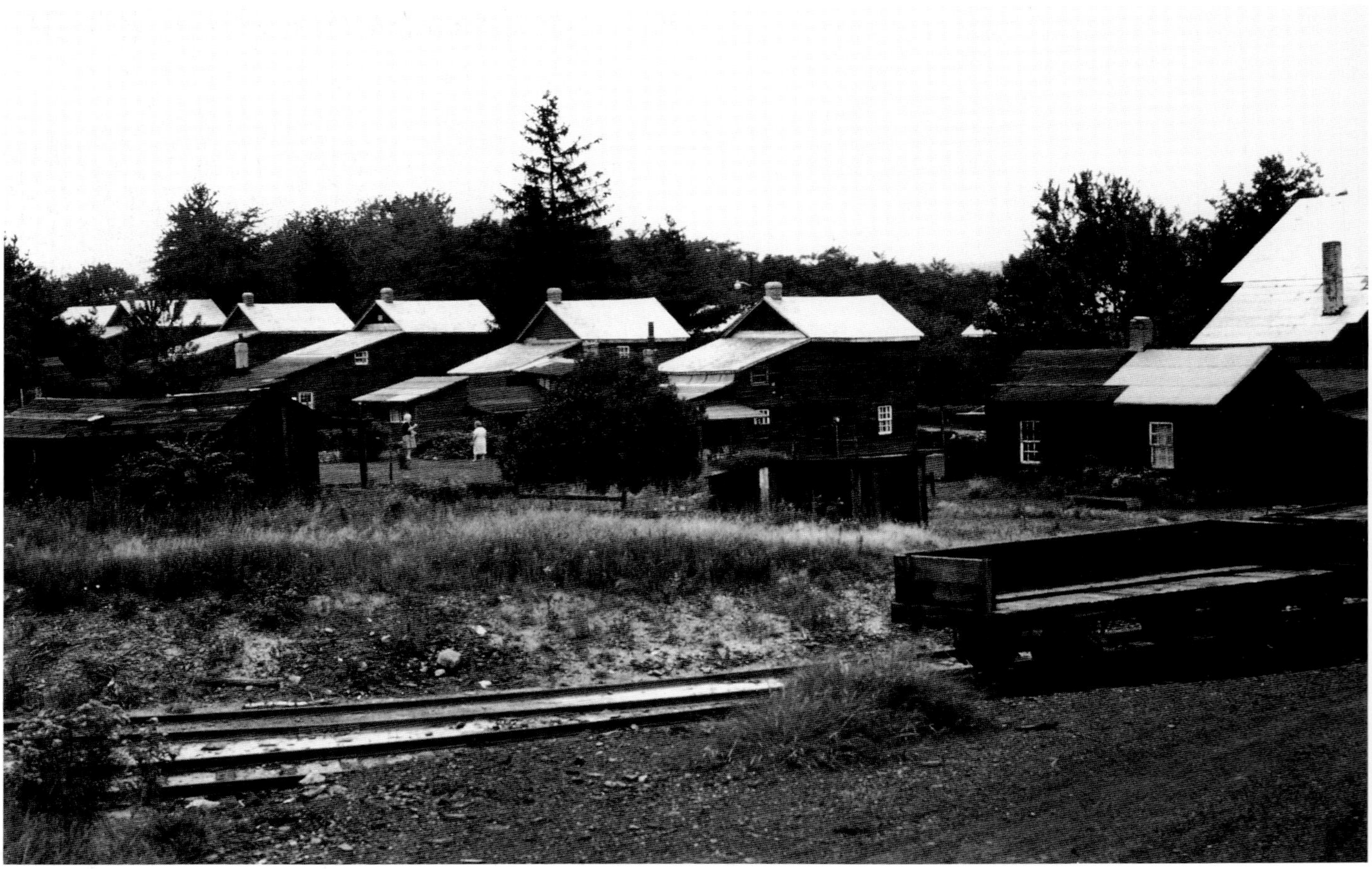

Eckley Mine Patch.

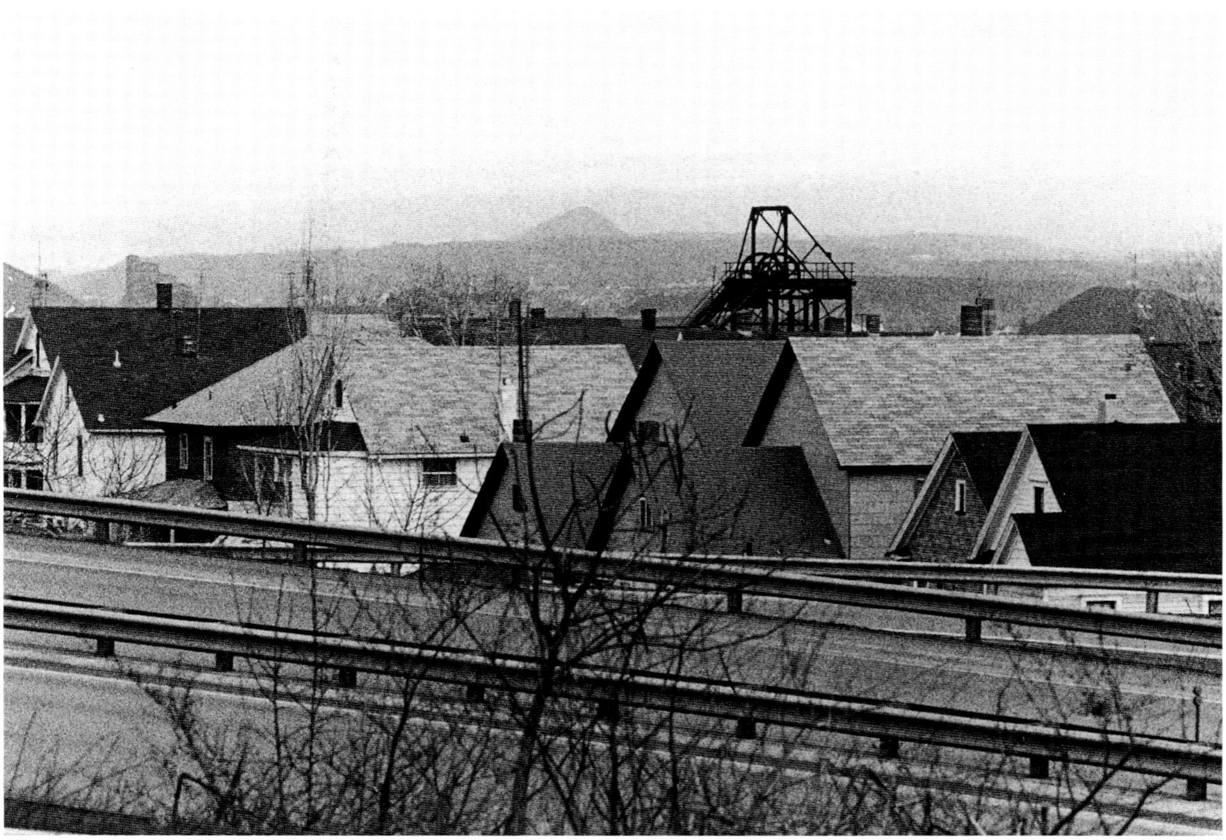

Eckley, a mine patch, has remained unchanged since the 1800s, a reason why it was chosen for the filming of *The Molly Maguires.*

"You took the clothes home over the weekend for washing. I don't know if anybody changed oftener than a week. In one day your clothes would be the same as they would in a week; full of dust and sweat."—Joe Sleyzak

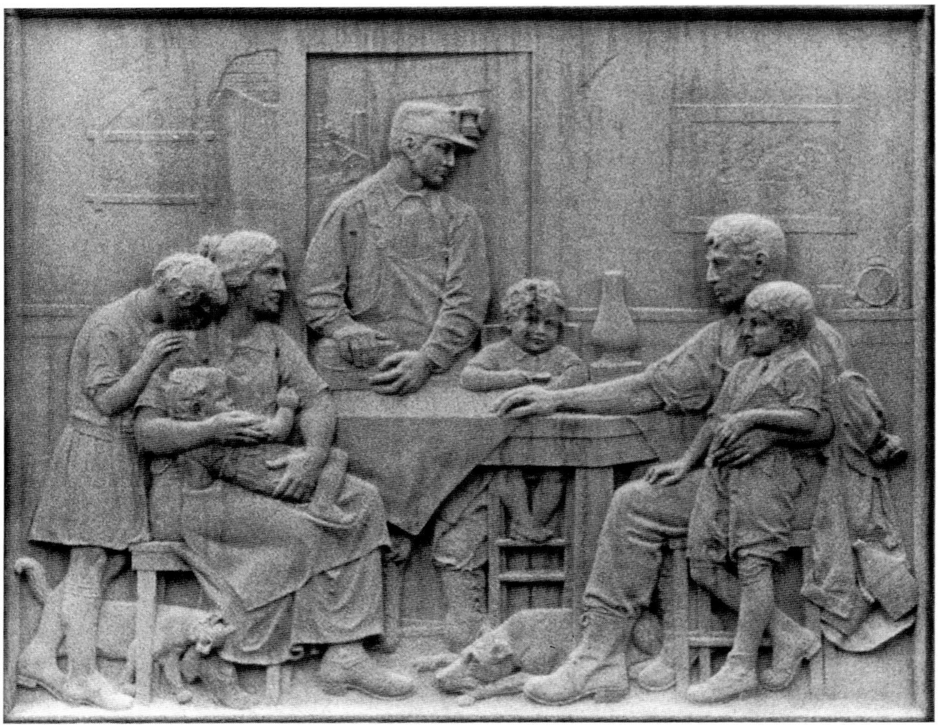

Anthracite Miner's Family, a bas-relief detail from the John Mitchell Monument, Scranton, 1924.

Official reports show that the foreign born and their children of the first generation furnish a large percentage of the labor in the anthracite fields: Lackawanna County with 58%; Luzerne County 58%; Schuylkill County 41%. The gradual replacement of Anglo-Saxon mine workers by Slavs has had a significant influence on the standard of living in the anthracite region. Early days of the industry labor . . . English, Welsh, Scotch, Irish mining customs and traditions were brought from England. In subsequent years this group was replaced by Slavs from Poland, Lithuania, European Russia, Hungary and Black Sea countries. Underground workers were Poles, Lithuanians, Russians. Italians and Americans worked in the breakers.—National Industrial Conference Board, *The Cost of Living Among Wage Earners: Anthracite Region of Pennsylvania,* 1922

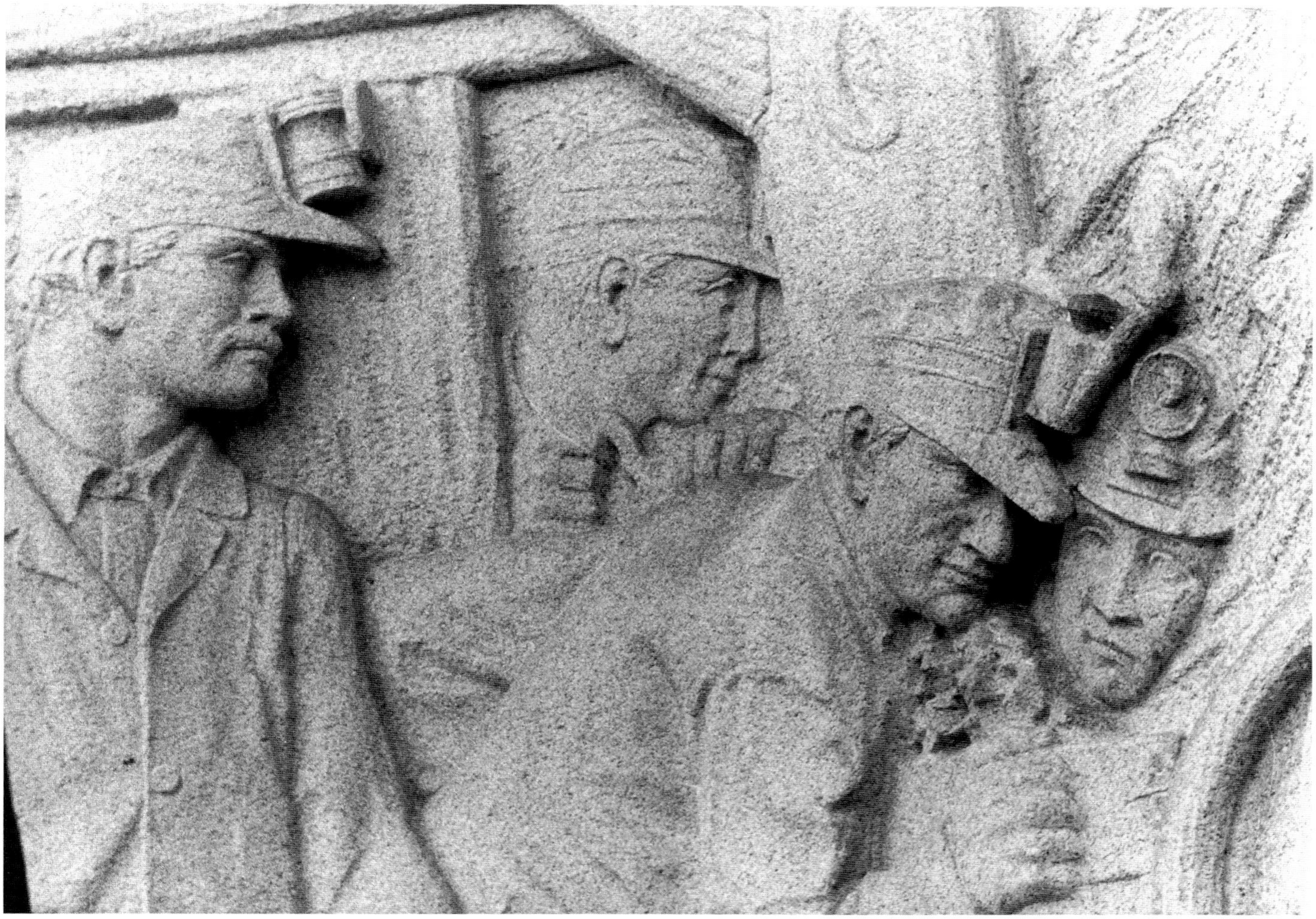

Bas-relief detail from the John Mitchell Monument, Scranton, 1924.

It's not Irish coal nor Polish coal,
Johnny Mitchell said.
The young one's blood on the winter slate,
No matter whose it is, stains red.
And when the women watch the small lads go,
Their lunch kettles touching the snow,
It's not Hungarian tears they shed
But tears that only mothers know.
—Madeline Mulveney Downry,
Down on the Coal Fields, 1902

"Mining has rarely been carried on except by pioneers or slaves. It has always been colored by the spirit of discovery and adventure. The pioneer seeks always to better his condition and hopes to reap rewards greater than he could by remaining among his fellows."—John L. Lewis, 1925, President of the United Mine Workers of America for many years

I sat down on the steps and surveyed the group of men. Miners. Broad shouldered laborers who walked with a slouch and stood with a stoop. Leathery work-worn faces of European cast, frequently grizzly with mustache and beard. Trousers and shirts earth-colored. In the half-light, they seemed strangely earthy with their hairy fists and faces and guttural voices.
—Anthony Bimba

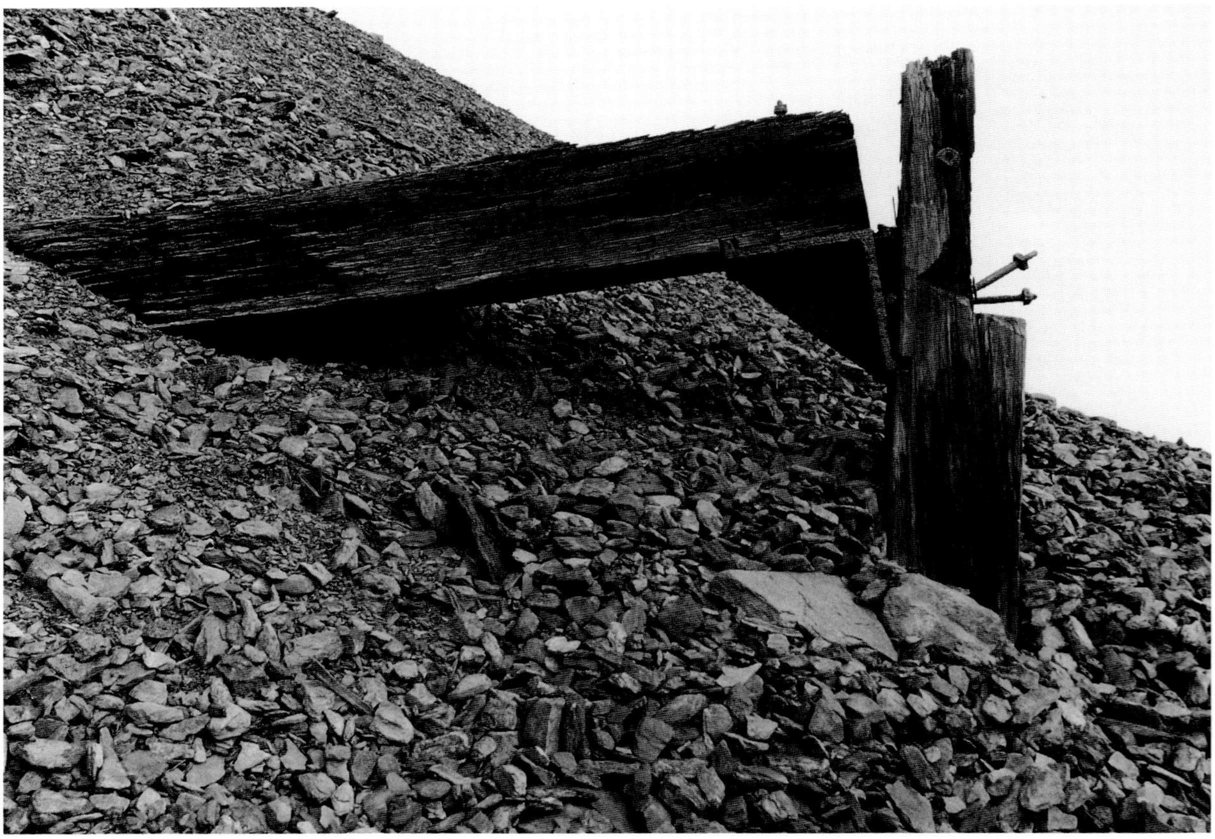

CULM BANK

Mountainous waste piles of slate and rock,
Black banks of coal refuse that stand
Stark and forbiddingly barren,
Permitting a scattering of low scrub and scrawny trees
With a few black dust covered leaves.
Mountainous black banks of coal refuse
Some turned copper-color from fires within
That burn on and on and on.
—The author

$5,000,000 was spent in the 1960s extinguishing nine-tenths of the internal fire in the Huber Culm Bank near Ashley.

"Before the union, before they were organized, if the man at the shaft didn't like the looks of your coal, it was dumped on one of these banks. Through the years there's a lot of coal in there. These culm banks are in our backyards. It's all that's left around here anymore. There were breakers all around us here at one time."—Joe Sleyzak

In 1912, there were 27 collieries and 33 mines operating within the city of Scranton.

Now at the lower end of town we used to have culm banks anywhere from seventy to a hundred feet high. We, as young boys in this town, used to go to the colliery and get long pieces of sheet iron. We would curl up the end like a toboggan, put a wire on the back of it, pull the sheet iron up to the top of the culm bank and then ride down on it.
—William W. Lewis, *Black Rock*

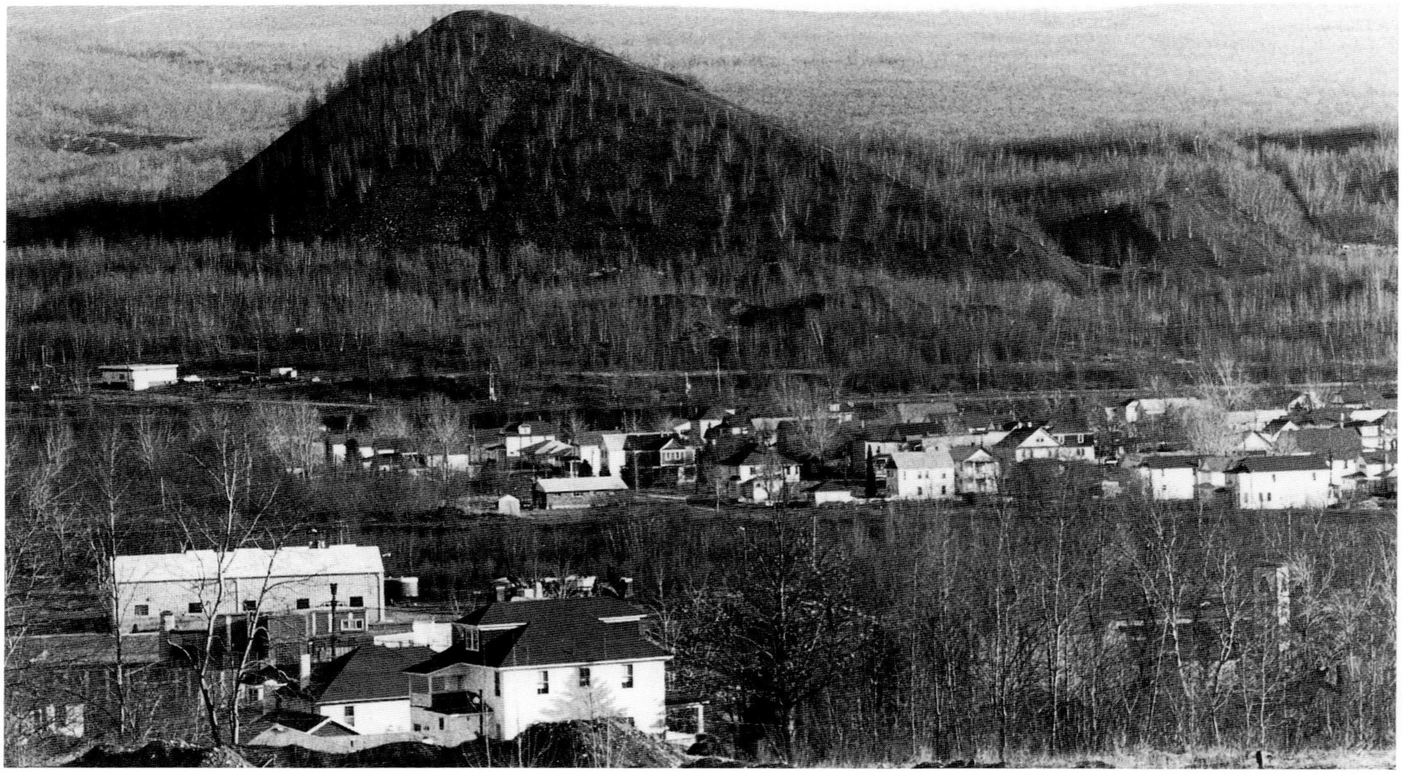

A culm bank old enough for trees to grow on it.

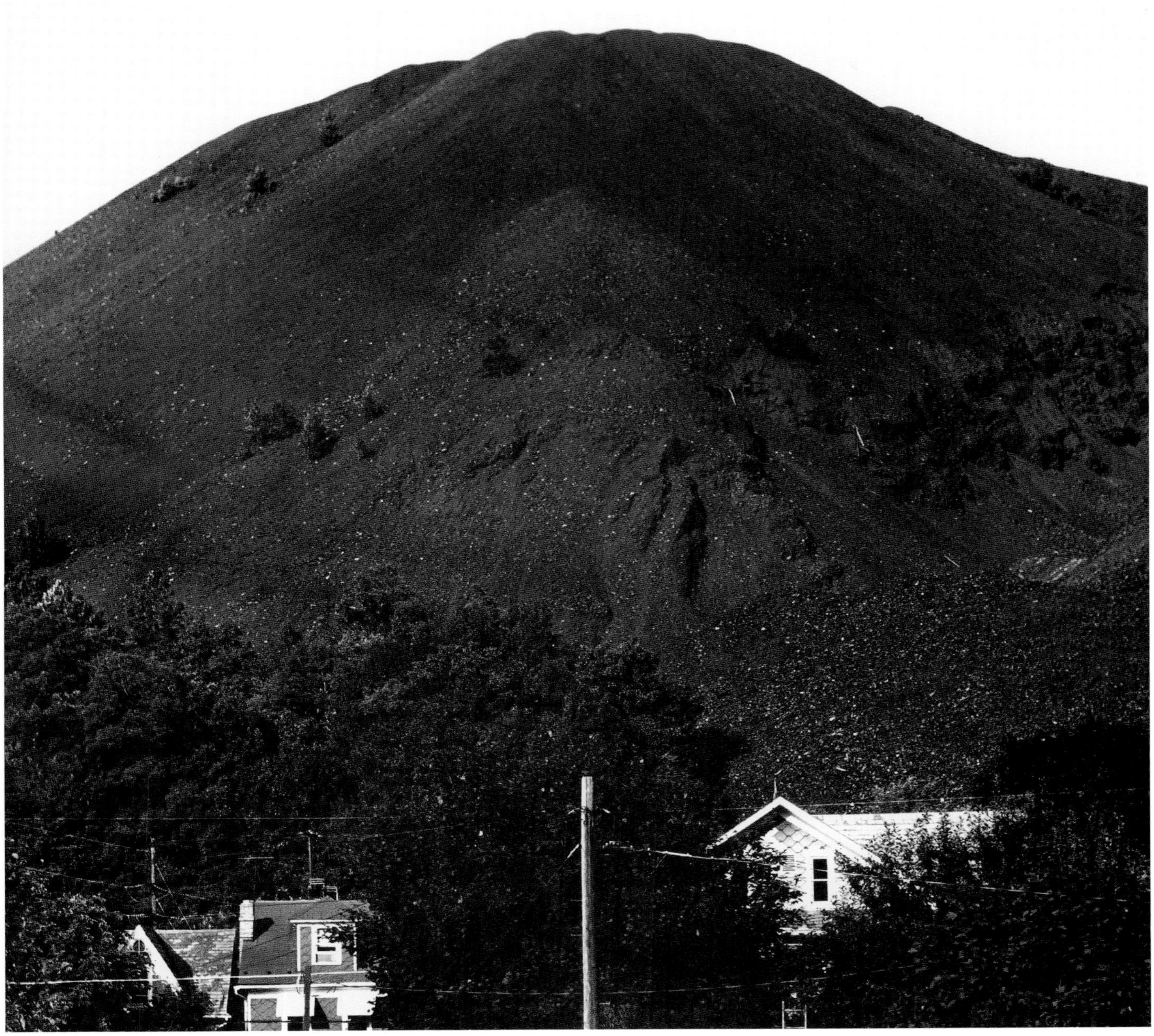

A massive culm bank in Scranton. Culm banks are neighbors of residential houses in coal communities.

CULM BANKS MAIM AND KILL

Destitute conditions drove widow Fadden to pick coal from a culm bank when a culm heap fell on her, breaking a leg.—*Scranton Republican,* October 31, 1877

Patrick Heffron was found dead on a burning culm bank near No. 11 breaker. It was presumed he suffocated from sulphur fumes.—*Scranton Republican,* October 11, 1877

William Thomas, 51, a colliery foreman for 16 years at Franklin Colliery near Wilkes-Barre, was smothered under a mass of culm. After discovering a fire in a culm pile, Thomas sent a work force to remove part of the side of a bank to find the seat of the fire so it might be extinguished. While directing the project, a vertical wall of culm fell on him. When workers reached him after 15 minutes of intensive digging, he was dead.
—*Scranton World,* February 16, 1864

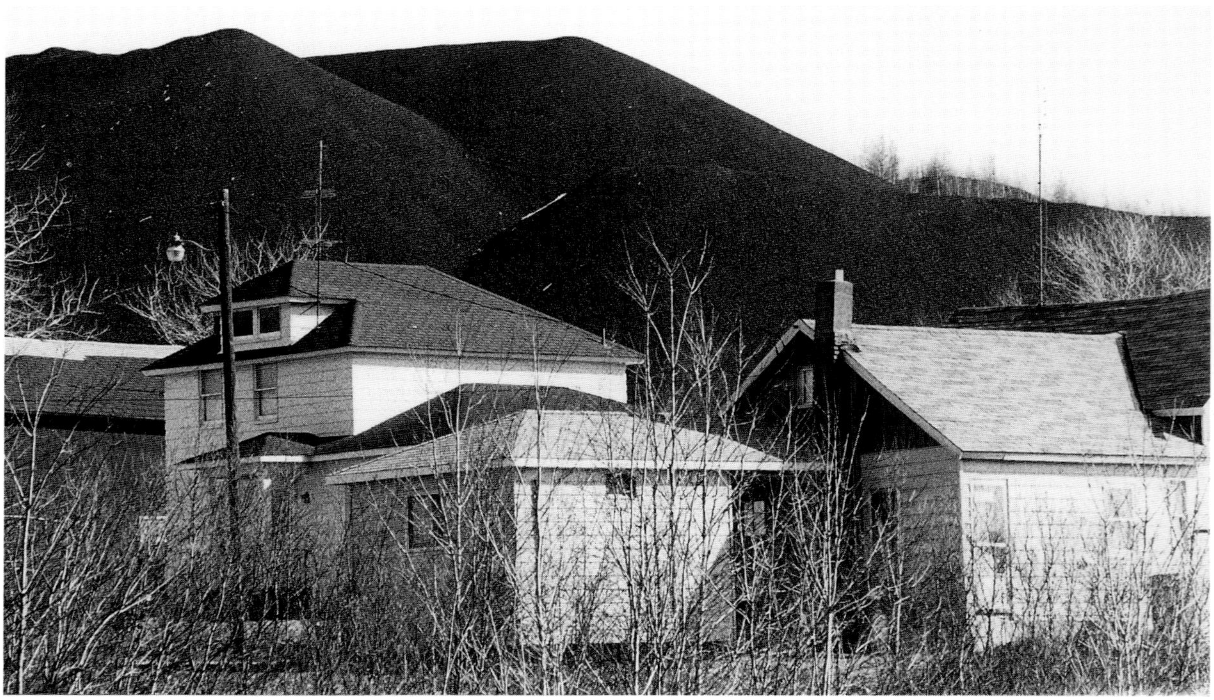

Residences in the shadow of a culm bank, typical throughout the anthracite area.

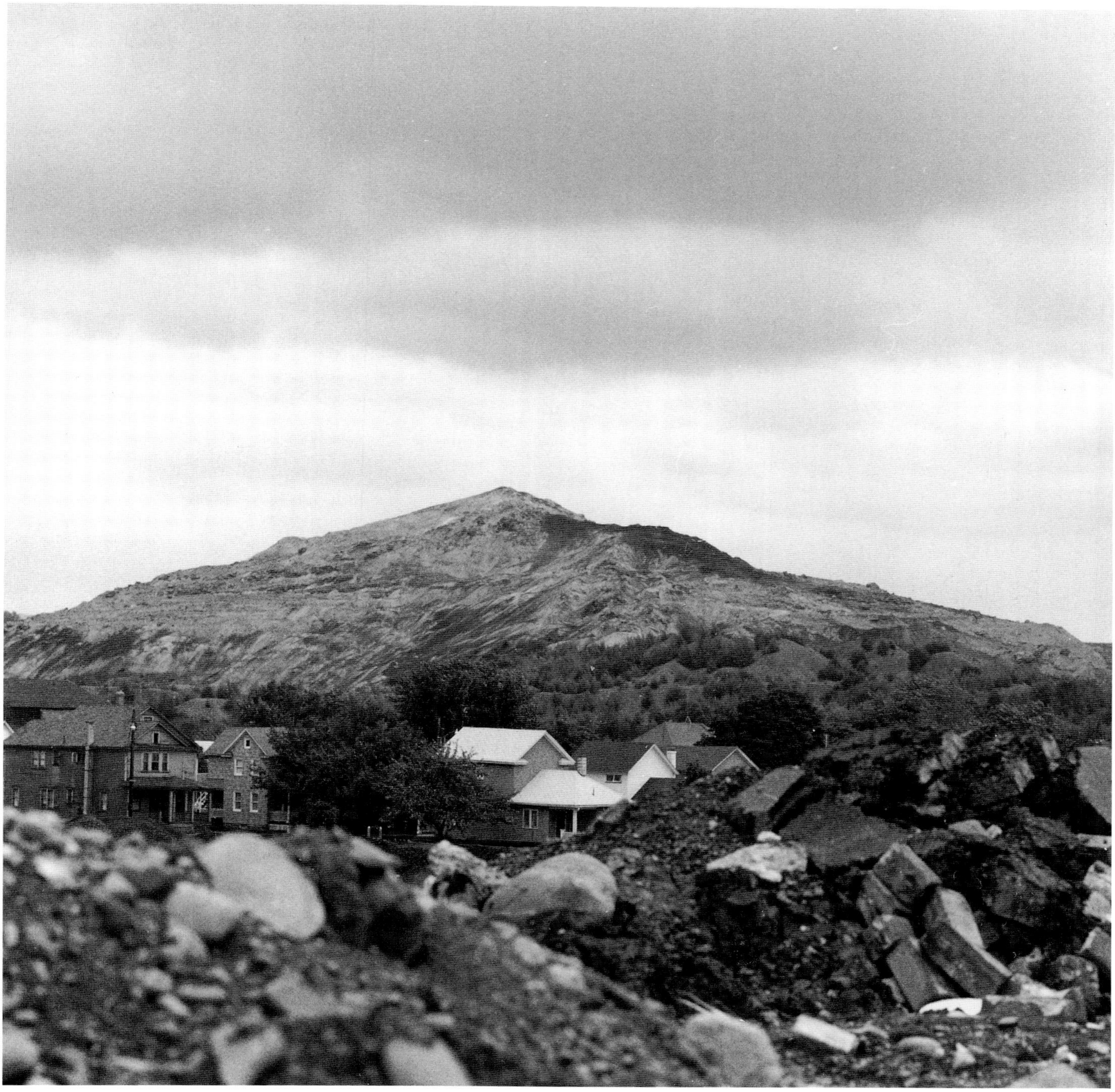

The irregular shape of Eddy Creek Culm Bank due to its gradual removal during the extinguishing of underground fires; Dickson City homes are in the foreground, 1972.

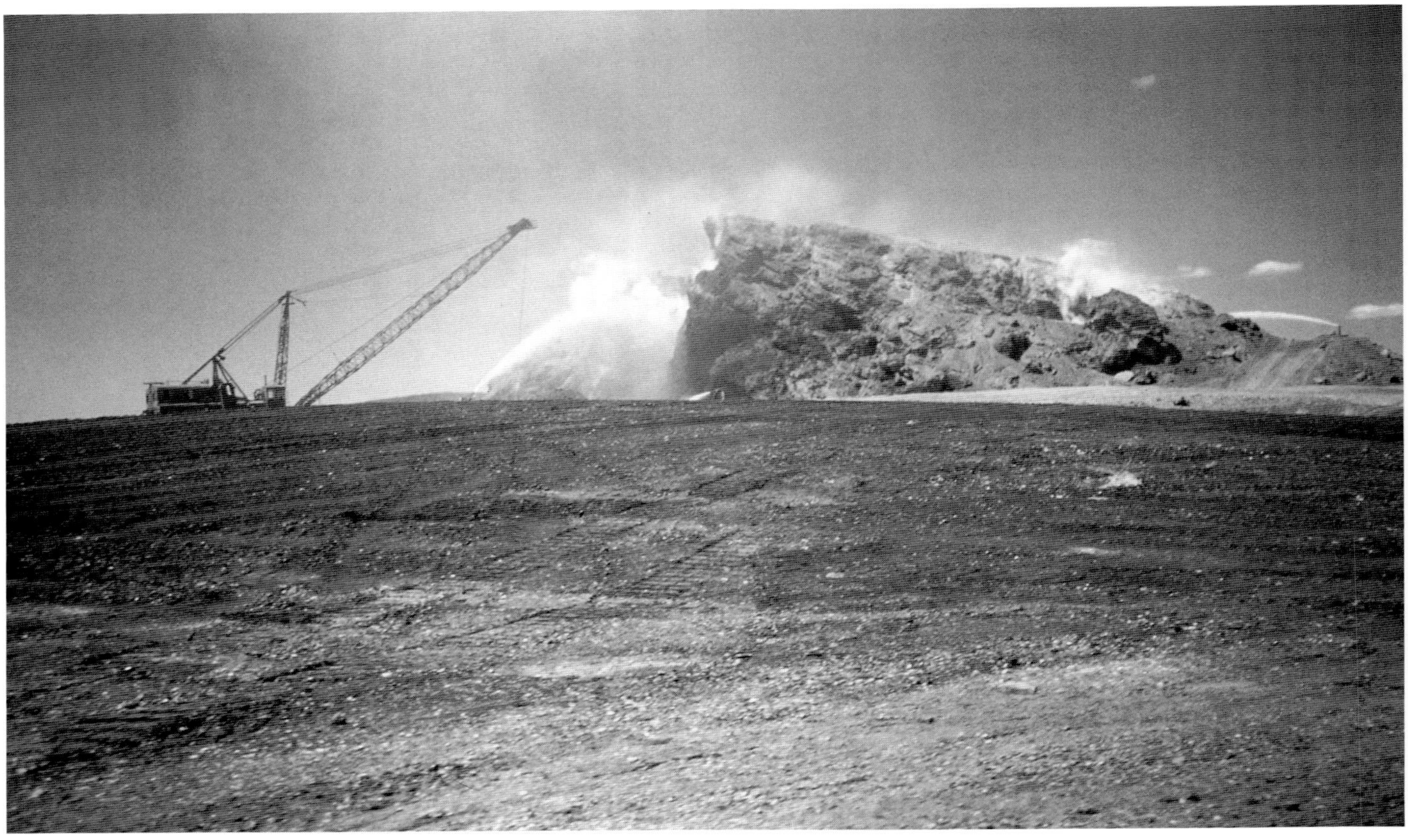

The fires in the long burning Baker Colliery Culm Bank were extinguished by removing the bank; Scranton-Taylor line, 1973.

Isaac Weil, 42, one of the few Jews employed in a regional mine, and Michael Marwell, 30, both single, were fatally burned at Ewen Colliery. A report noted that the accident occurred when Weil and Marwell threw water on some burning culm causing the resulting steam to explode. —*Scranton Times,* April 7, 1906

Two small boys, George Wilkes and Harry Scott, while running across the Old No. 4 Delaware and Hudson Culm Dump, came across the lifeless body of a man on the culm. The man's face had been eaten away, the right side of his body was badly burned and his left hand was almost burned off from the heat generated by the burning dump.

Finding of this body caused much excitement on Welsh Hill, it having been the second one discovered there within a two-month period. It was thought that since the culm bank had been on fire for several years, it was a place of refuge for lonely men in quest of a place to have a warm night's sleep.—*Scranton Times,* March 6, 1903

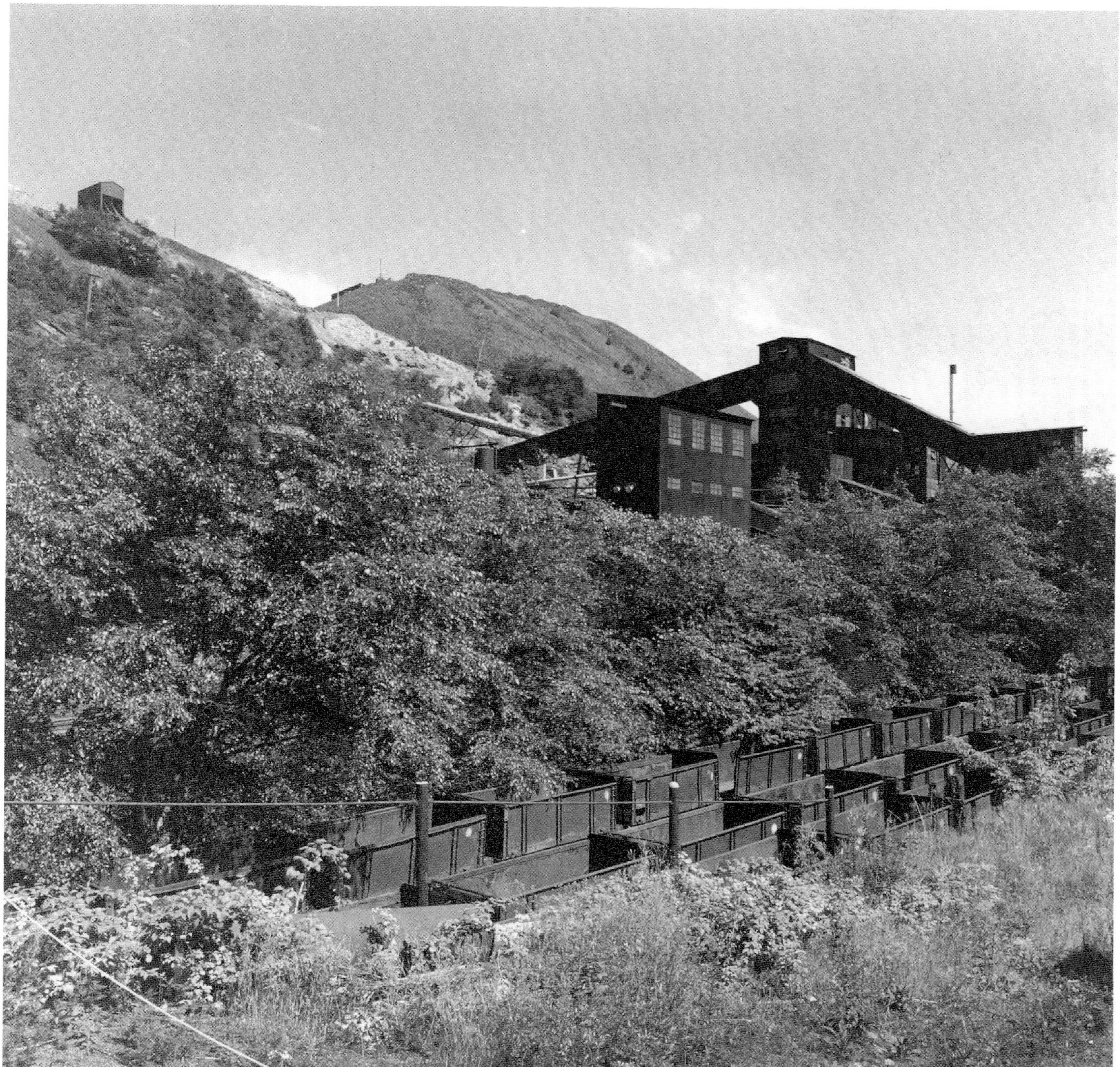

Deep underground fires smolder in the culm bank high above Glen Burn Colliery, Shamokin, 1972.

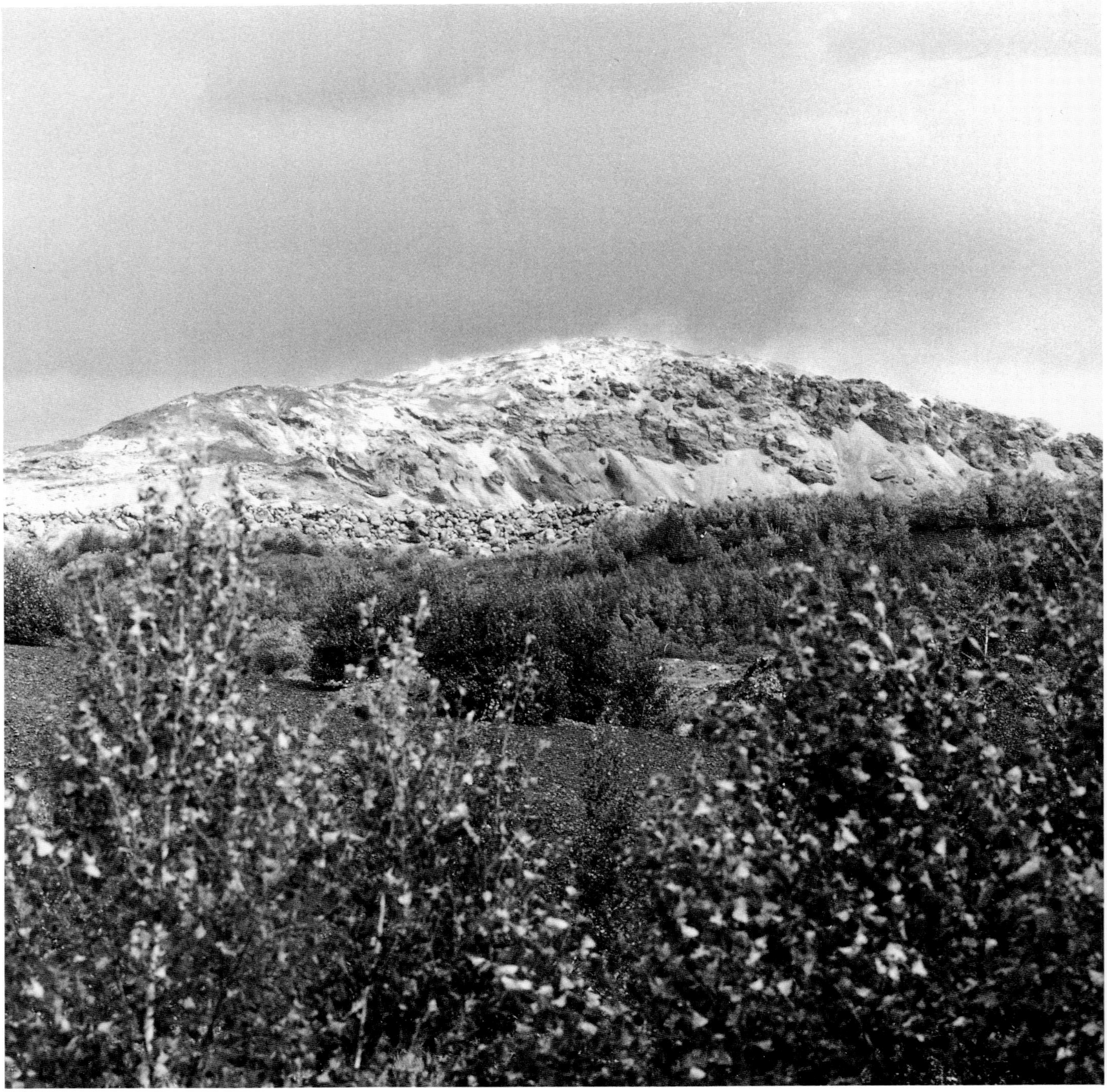

A burning culm bank.

I looked away at the hills in the distance. It struck me that a group of them was the same dark color gray as the slag heap.

"Are those hills over there? They ain't got no trees."

"Them's slag dumps."

"Slag dumps! Why, they're as big as hills. How long does it take to pile up a hill like that?"

"Maybe seventeen, maybe twenty year."

"You shovel all that?"

"Yeah."

"You get paid for it?"

"Naw, nottin'. We only get paid for coal."

—Laura Gilfillan, *I Went to Pit College*

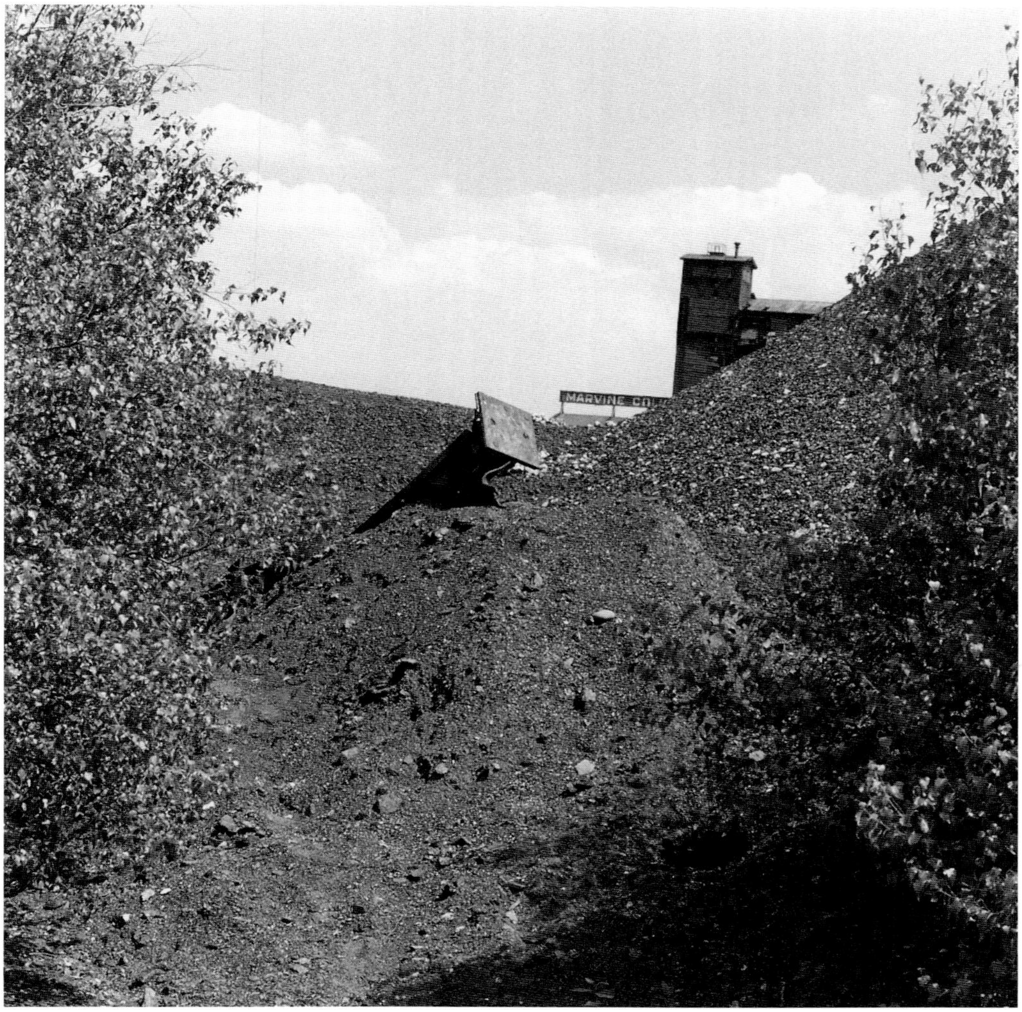

Marvin Colliery, Scranton, 1971.

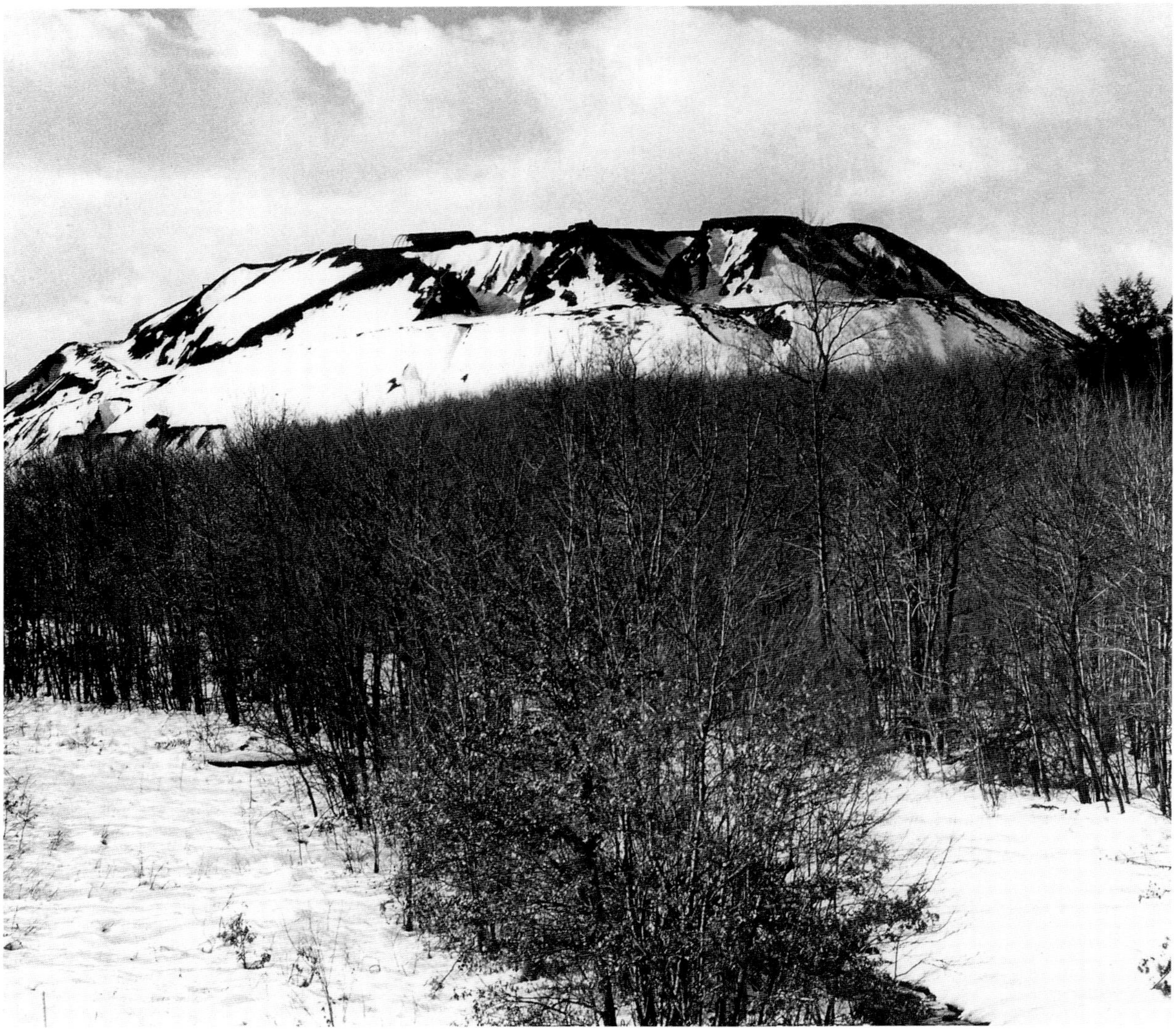

A culm bank near Mahanoy City. Coal is being removed for fuel.

I looked out the window. Rolling mists enveloped the hilltops in the clouds of greenish milkiness, drifted and shifted in the valley, filling it like a cup. A mine tipple rose black among the trees. Down there, the narrow roadway of cinders which led into the village. And everywhere the smell of coal.—Laura Gilfillan, *I Went to Pit College*

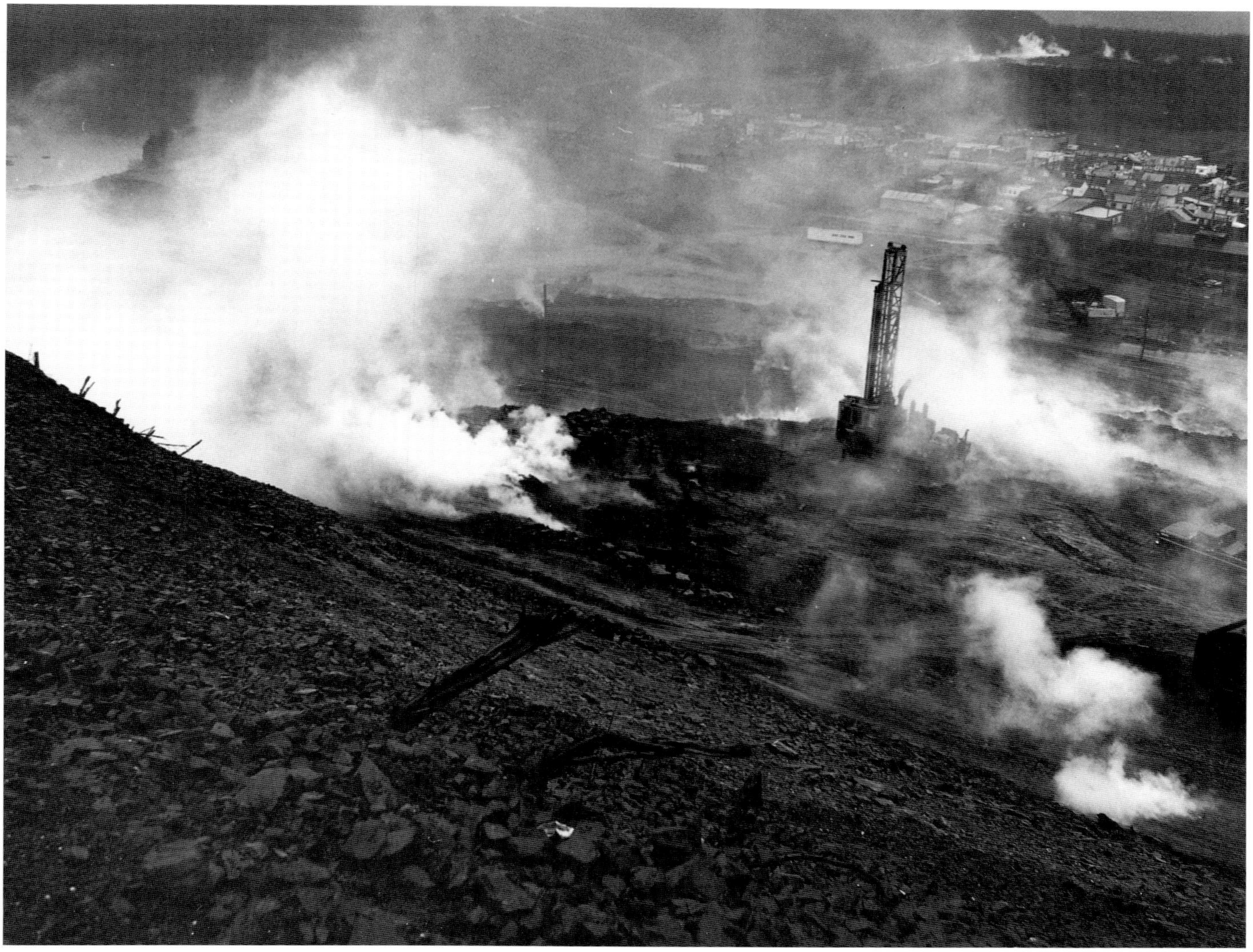

Work in progress to extinguish underground mine fires in a mountain overlooking Shenandoah.

Yren stood solemnly looking at me.

"Don't never wash your hair with that there water."

"Why not?"

"It'll all come out. It's sulphur."

"How do you wash your hair?"

"When my mother died, Grandma catched rain water an' washed my hair to go to the funeral."

—Laura Gilfillan, *I Went to Pit College*

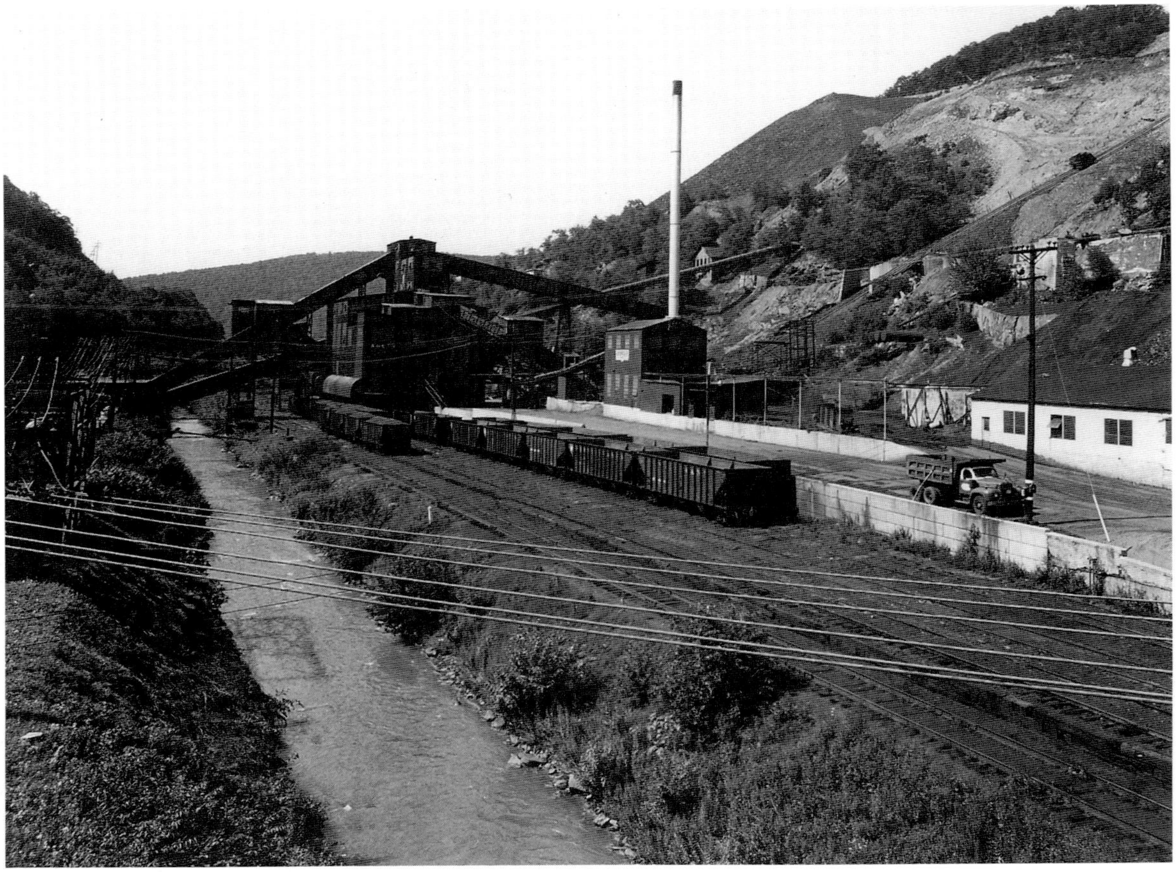

A coal mine in a mountainside feeds directly into the colliery. Water from the mine operation pollutes the stream.

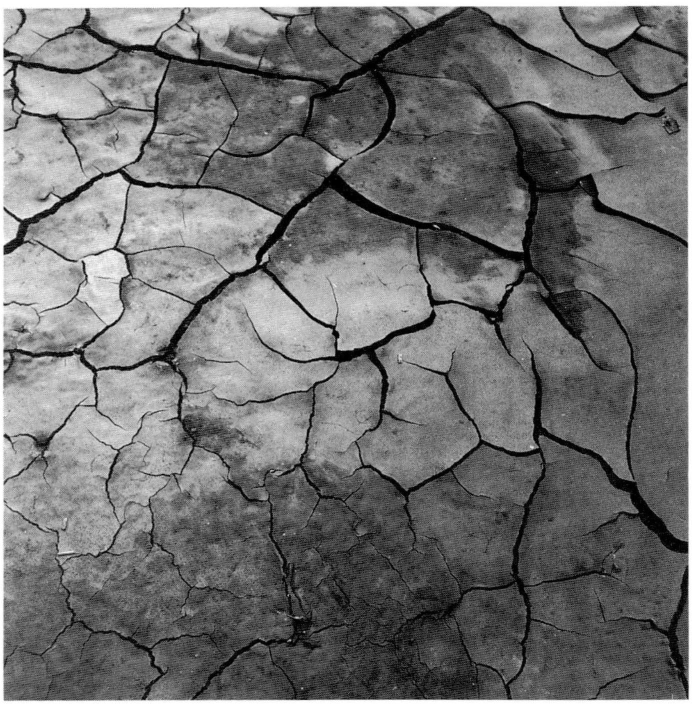

Water from washing the coal carries coal dust to a drainage area. The resulting silt dries with a dark gray color.

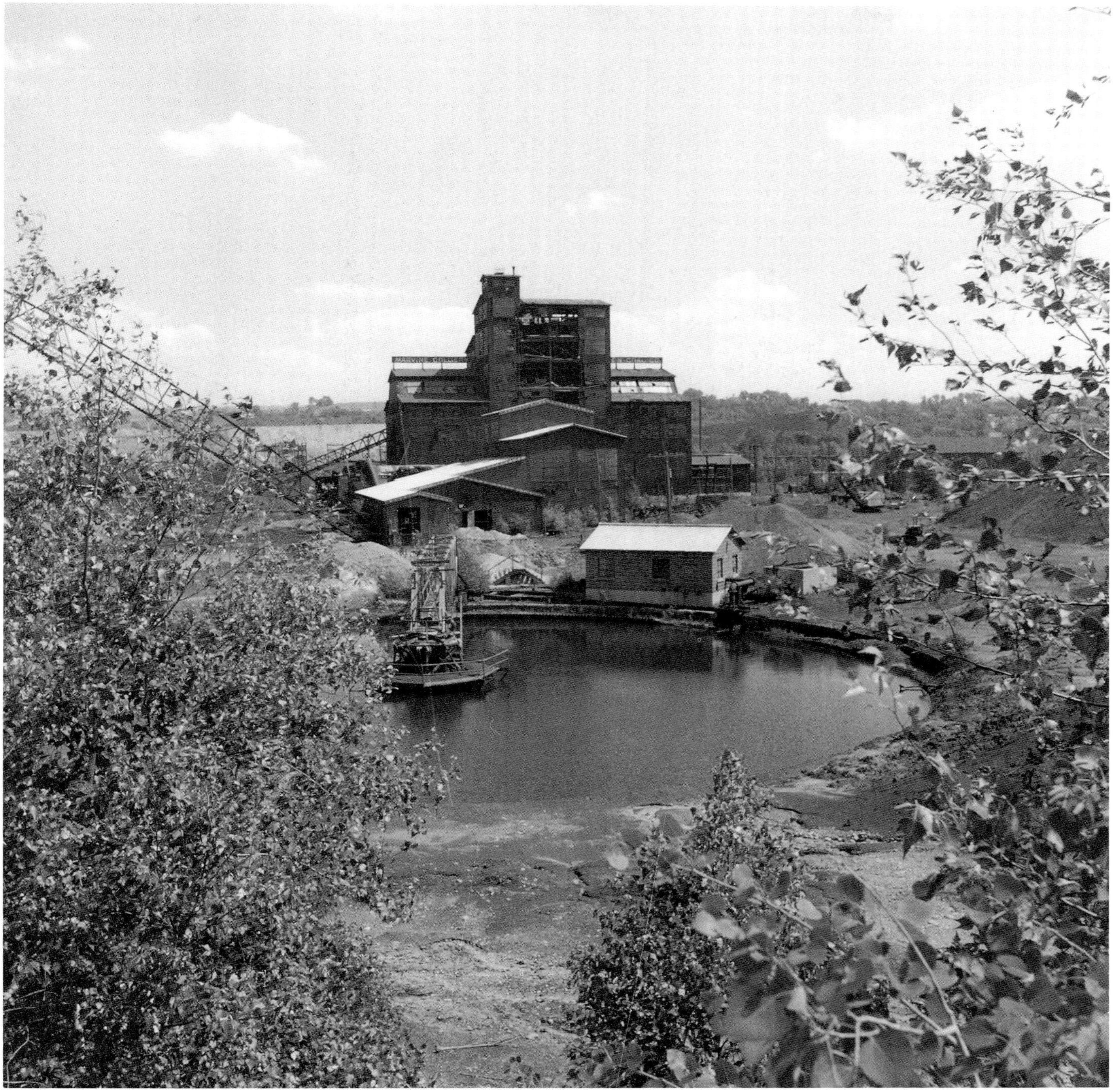

Marvin Colliery, Scranton, 1971.

During the thirteen years between 1862 and 1875, there were 142 unsolved homicides and 212 felonious assaults in Schuylkill County. Many of the victims were mine superintendents, foremen, or colliery supervisors, who had incurred the wrath of a Molly Maguire or a friend of a Molly.

There never was any doubt that the operators of coal mines drove the hardest possible bargain; miners worked long hours, lived in squalor, died hungry, and their children lived without hope. Since the owners lived a hundred or more miles from their mines, they regarded the Molly Maguires with loathing rather than fear.

Non-union towns are towns of fear, declared a card circulated during the 1922 strike:

Non-Union Miners Have Lived In Fear!

- Fear of the Boss
- Fear of the Spies and Spotters
- Fear of Gunmen and Coal and Iron Police
- Fear of Anti-Union Civil Authorities
- Fear of the Blacklist
- Fear of Evictions

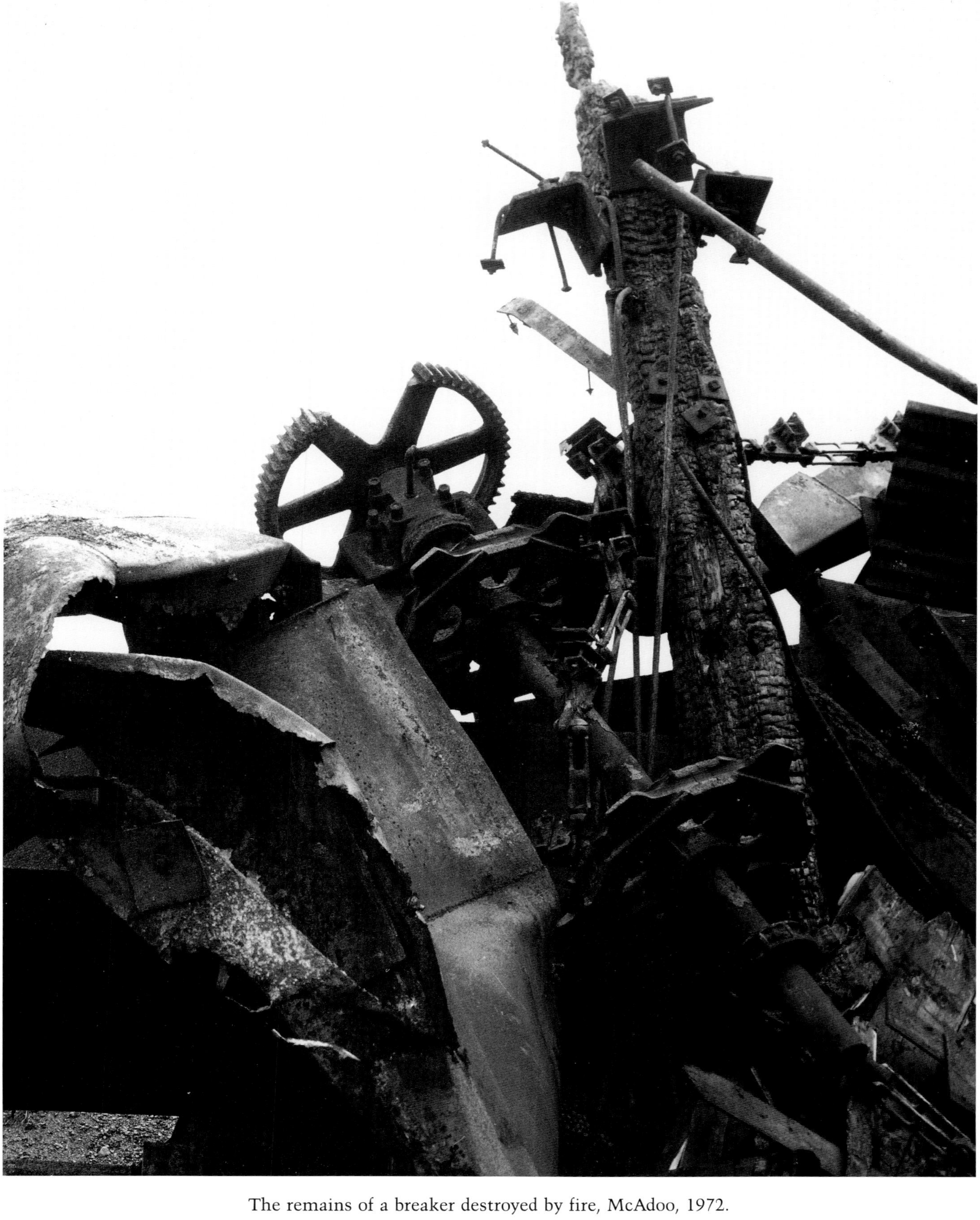

The remains of a breaker destroyed by fire, McAdoo, 1972.

Some say that the name Molly Maguires was designated by the coal interests to Hibernians in the miners' union who put up a determined struggle against conditions of starvation and peonage in the coal fields. The Mollies were the miners' leaders, at the dawn of a new large-scale industry, in the first major class battle in American labor history. The first anthracite strike was on July 8, 1842.

What terrified these absentee owners was the emergence of labor unions; these they tried to crush using any means they could, including use of their own private armies, which operated without regard for the Fourth Amendment.

From 1850 until 1877, three entire Pennsylvania counties and large portions of two more, with a total population of more than a quarter of a million, were without effective law and order in the face of an increasingly savage onslaught against on-the-spot representatives of absentee London, Boston, New York, and Philadelphia mine owners.
—Arthur H. Lewis, *Lament for the Molly Maguires,* 1964

DUNMORE MINE FIRE

PORT GRIFFITH SHAFT Number of victims estimated from twenty-five to fifty. First miner brought to surface: "Mine is a seething furnace; the roof is falling; coal is on fire; timbering in fire zone has been completely eaten away; belief that all in the mine have perished." Fire started at 10 AM 750 feet below the surface. When the fire started at ten o'clock this morning, he ran down the gangway to give the alarm. Before he had gone far, he realized the danger, and as the smoke was fast filling the workings, he was forced to retreat. He passed the worst place on his way back to the foot of the shaft, but had not gone far when he was overcome by the smoke. He was found some little time afterwards by rescuers and taken to the hospital in the mines, where they worked over him for hours and so critical was his condition that it was not until late this afternoon that they could bring him to the surface. Carried from the mine carriage to the office, where he was propped up in a big chair packed with blankets, he was able to tell in some detail of the horrors of the tragedy underground. In his opinion, there was absolutely no hope of any of the entombed being alive. "They perished hours ago," he said. "Nobody could live many minutes in such a place. The workings are filled with smoke, and there is no way of the entombed men getting air."

There were 300 men in the underground workings when the alarm was given. All but forty or fifty got out. Men who got out of the mine at 1:30 o'clock told a Times man that there is very grave doubts of the hemmed-in miners getting out alive, as they are trapped behind the wall of fire and smoke at the slope. Added to this is the danger from gas as the fire is setting off pockets of the poisonous stuff.

Nothing of the fire is visible on the surface.—*Scranton Times,* April 7, 1911

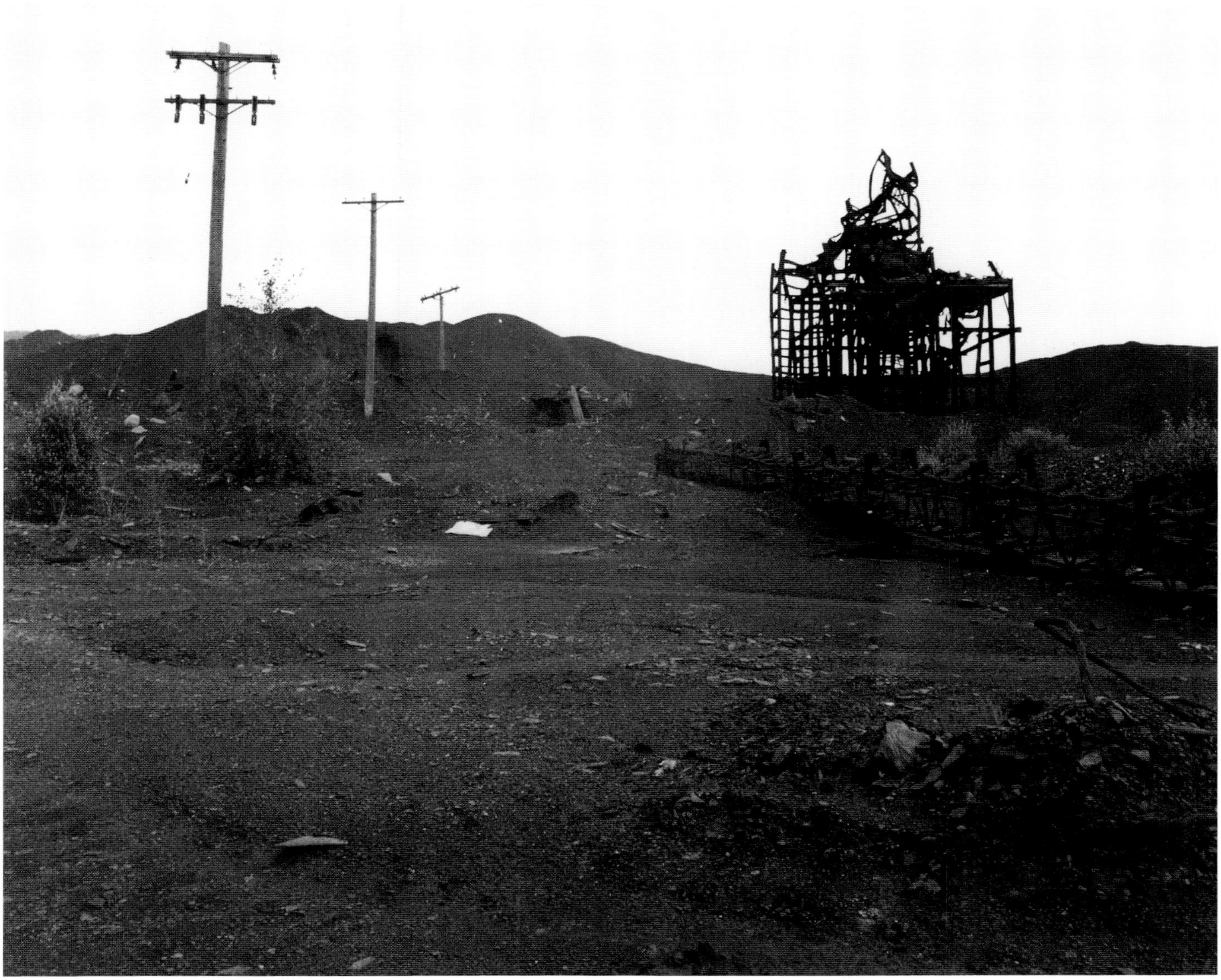

Old Forge, 1971.

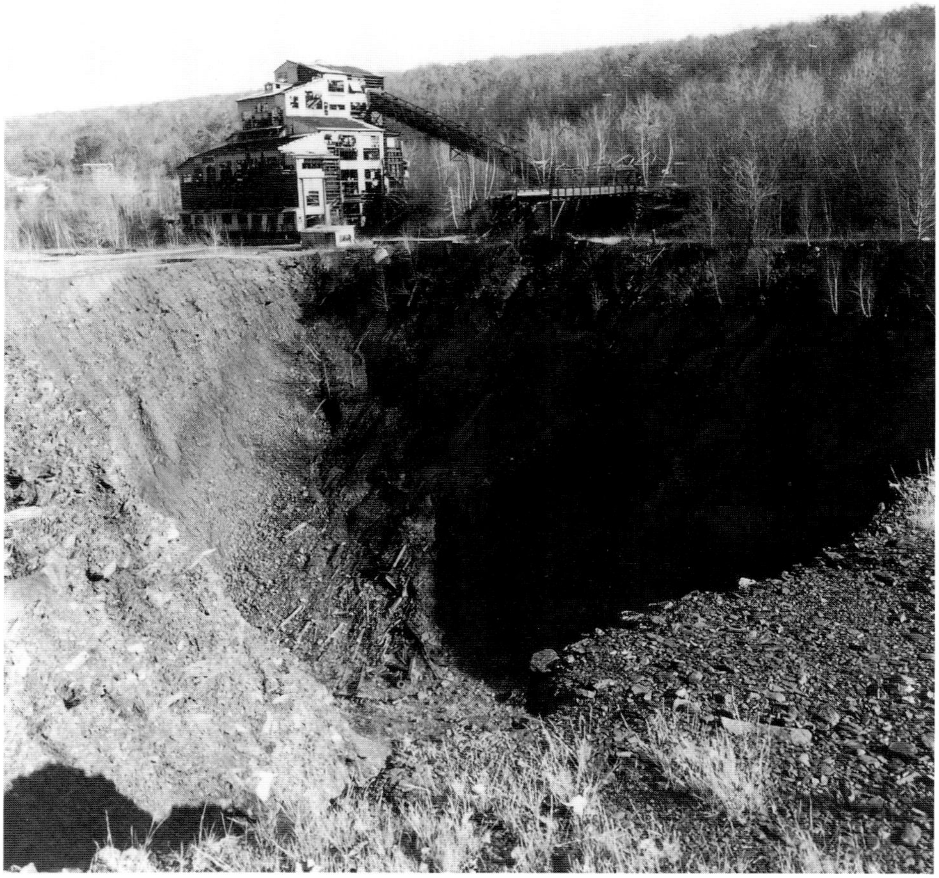

St. Clair Colliery after strip mining, 1972.

In memory of Spike Morgans, a fellow miner.
Death came along with solemn tread:
He smiled at me, took Spike instead.
—George W. Bowen, *Diamonds of the Mines,* 1928

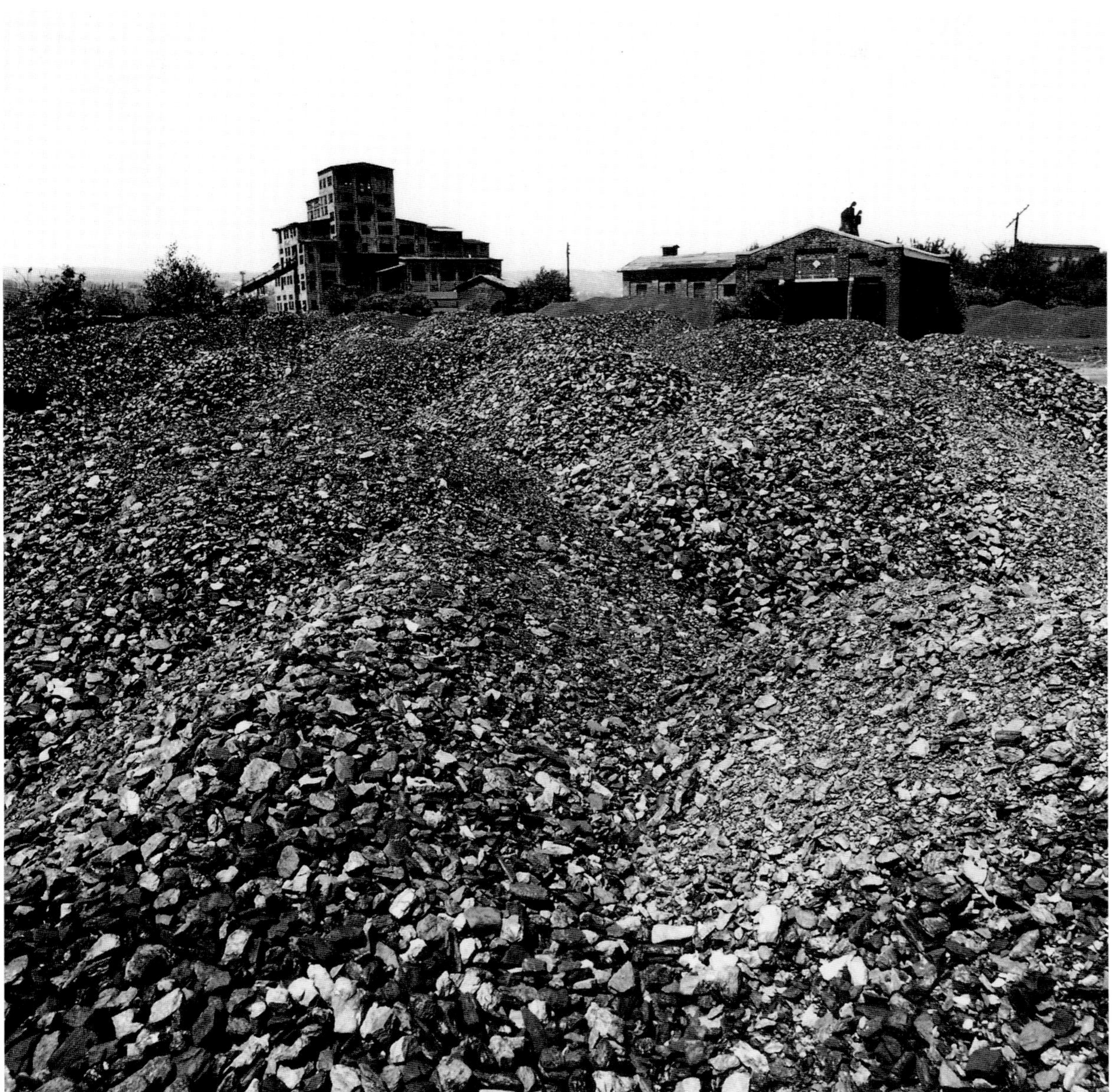

The Moffat Coal Colliery, Taylor, 1972.

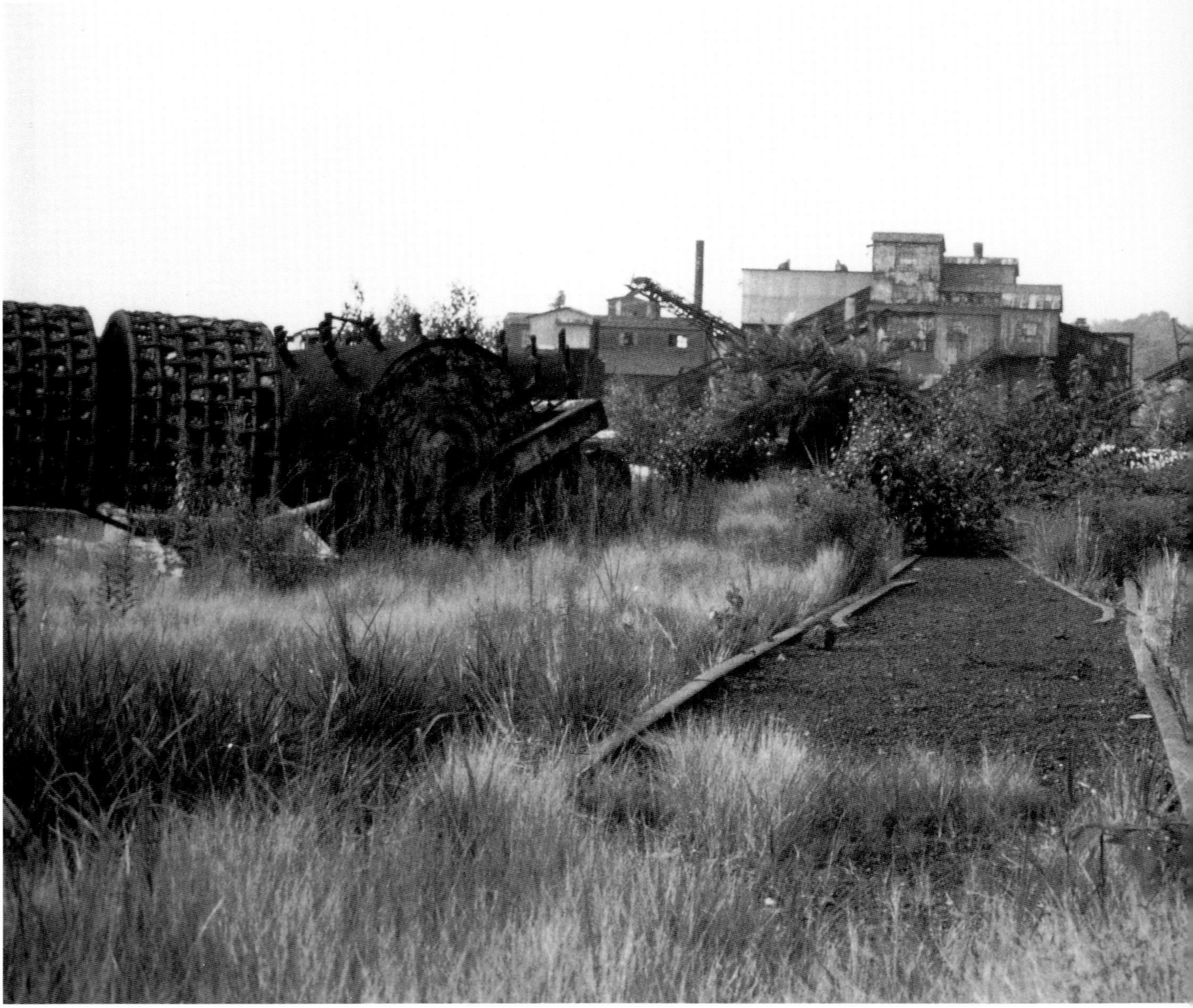

Franklin Coal Colliery, Wilkes-Barre, 1972.

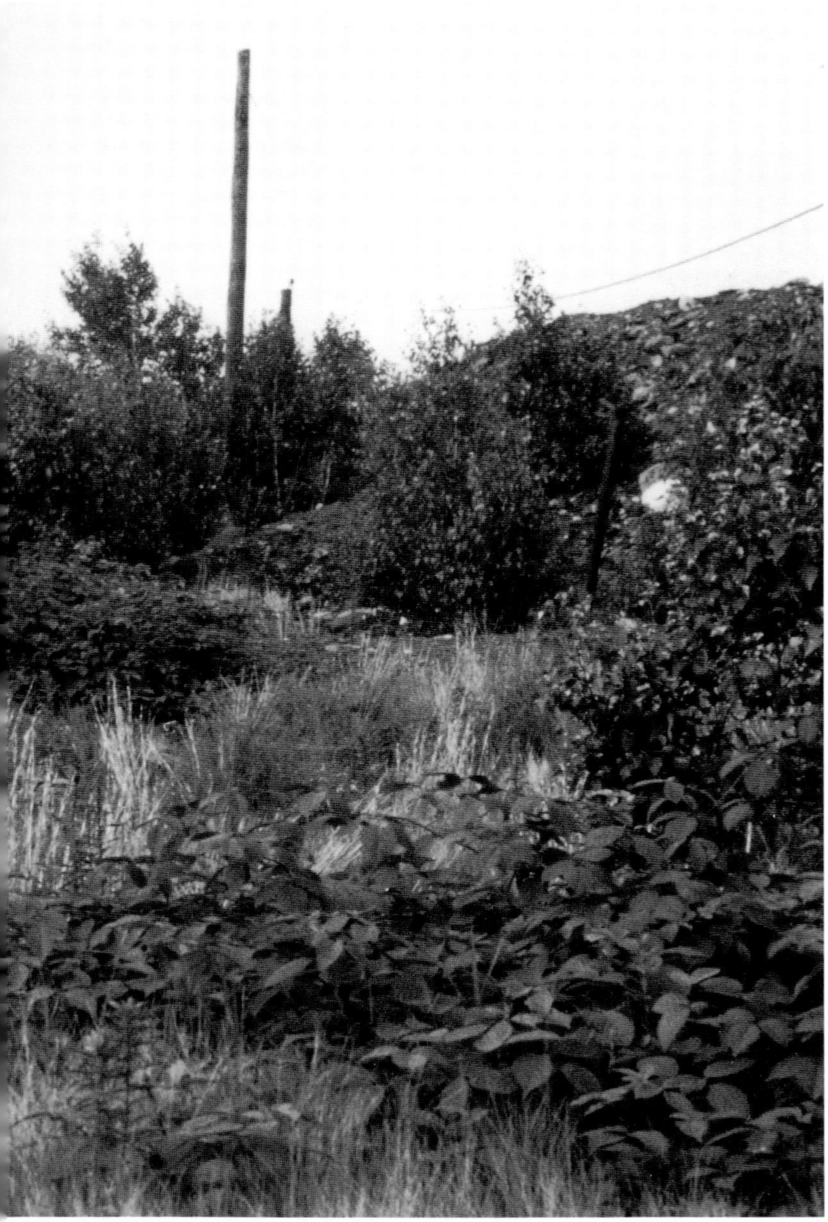

"Killed, said ye. i' the mine top fell?"

"Yes, an' struck him on the head; they said he didn't know what killed him."

—Homer Greene, *The Blind Brother: A Story of the Pennsylvania Coal Mines,* 1887

Good Christians all, both great and small, I pray ye lend an ear,
And listen with attention while the truth I will declare;
When you hear this lamentation, it will cause ye to weep and wail,
About the suffocation in the mines of Avondale.
On the sixth day of September, eighteen hundred and sixty-nine,
Those miners all then got a call to go to work in the mine;
But little did they think that death would gloom the vale
Before they would return again from the mine of Avondale.
The women and the children, their hearts were filled with joy,
To see the men go to work again, and likewise every boy;
But a dismal sight in broad daylight soon made them all turn pale
When they saw the breaker burning o'er the mines of Avondale.
But all in vain, there was no hope one single soul to save.
For there is no second outlet from the subterranean cave.
No pen can write the awful fright and horror that did prevail
Among those dying victims in the mines of Avondale.
Now to conclude and make an end, their number I'll pen down;
One hundred and ten of brave men were smothered underground;
They're in their graves 'til the last day, their widows may bewail
And the orphans' cries, they rend the skies, all 'round through
Avondale.

—James Fox, "The Avondale Mine Disaster," 1927

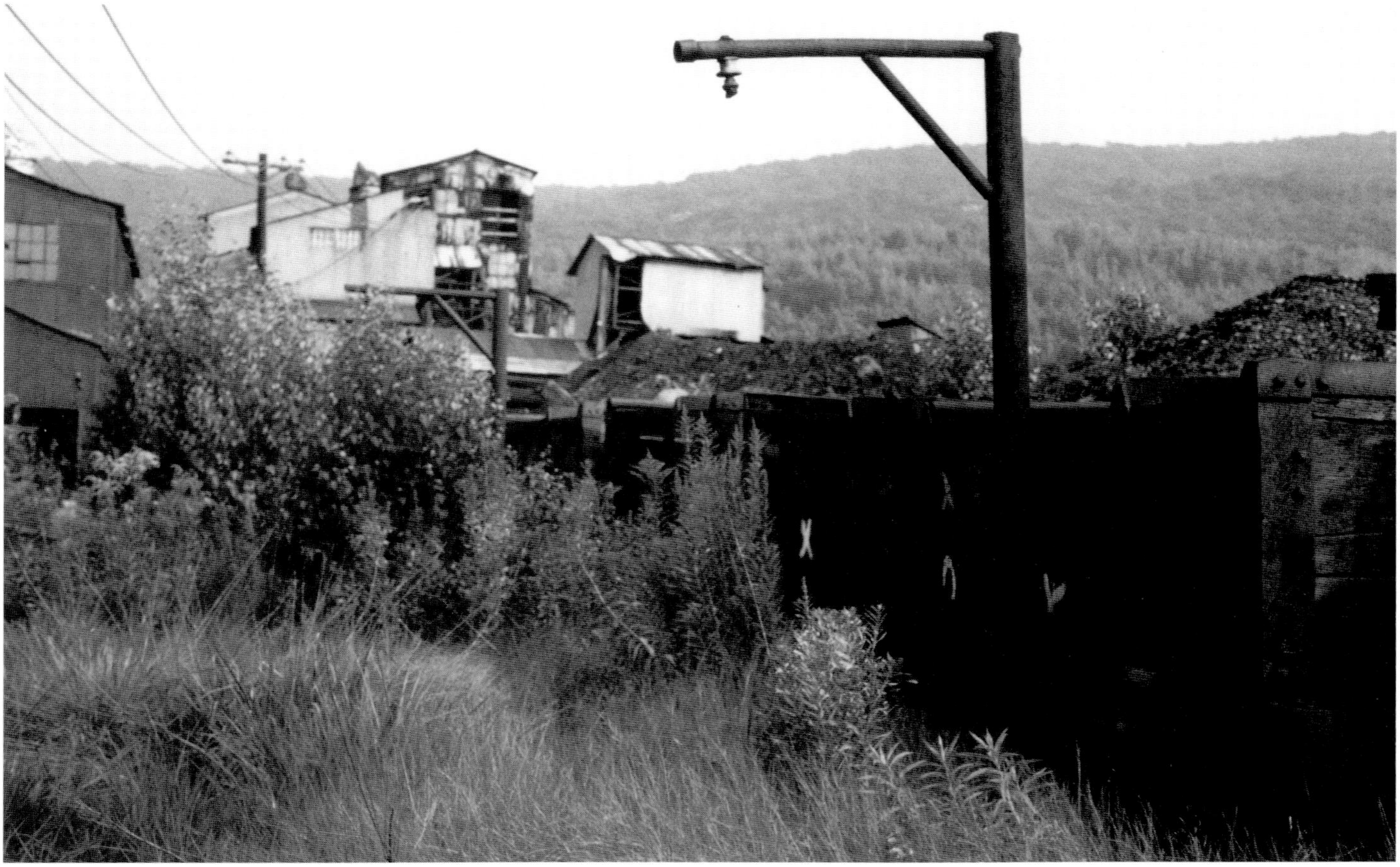

When the Franklin Colliery closed, many coal cars were left standing loaded with coal; Wilkes-Barre, 1973.

TWIN SHAFT DISASTER The Twin Shaft Disaster occurred at 3 A.M., Sunday, June 28, 1896, taking a toll of 58 lives, men and boys. This horrendous accident, a roof fall, occurred 500 feet underground and 1,500 feet from the front of the shaft, built in 1856. Output was 1,500 tons per day. Most of the men caught in this major catastrophe were family men who were survived by 31 widows, 101 orphans, a number of dependents, and a horde of sorrowing relatives and friends. Practically all the boys and men who died in this mine accident were of Irish and Lithuanian birth or descent. None of the bodies were ever recovered.

On June 28, 1896, fifty-eight men were killed in a massive cave-in of rock and coal here, in the Newton Coal Companies Twin Shaft Colliery. An investigative commission appointed by the Governor, reported on September 25. Although its safety recommendations would often be ignored, the disaster was a factor that led to a stronger unionization of this region under John Mitchell after 1900.
—State roadside marker near Pittston

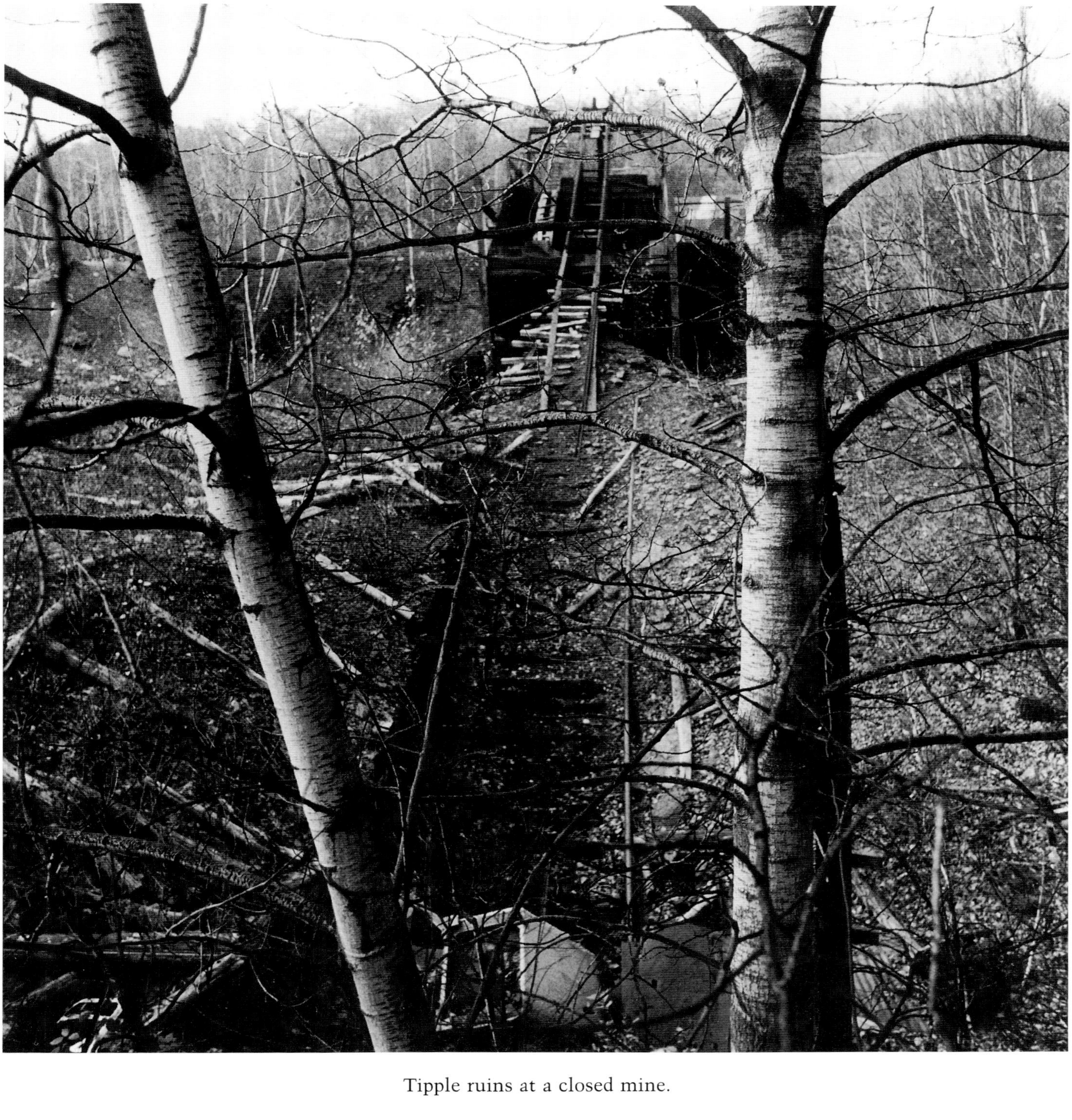

Tipple ruins at a closed mine.

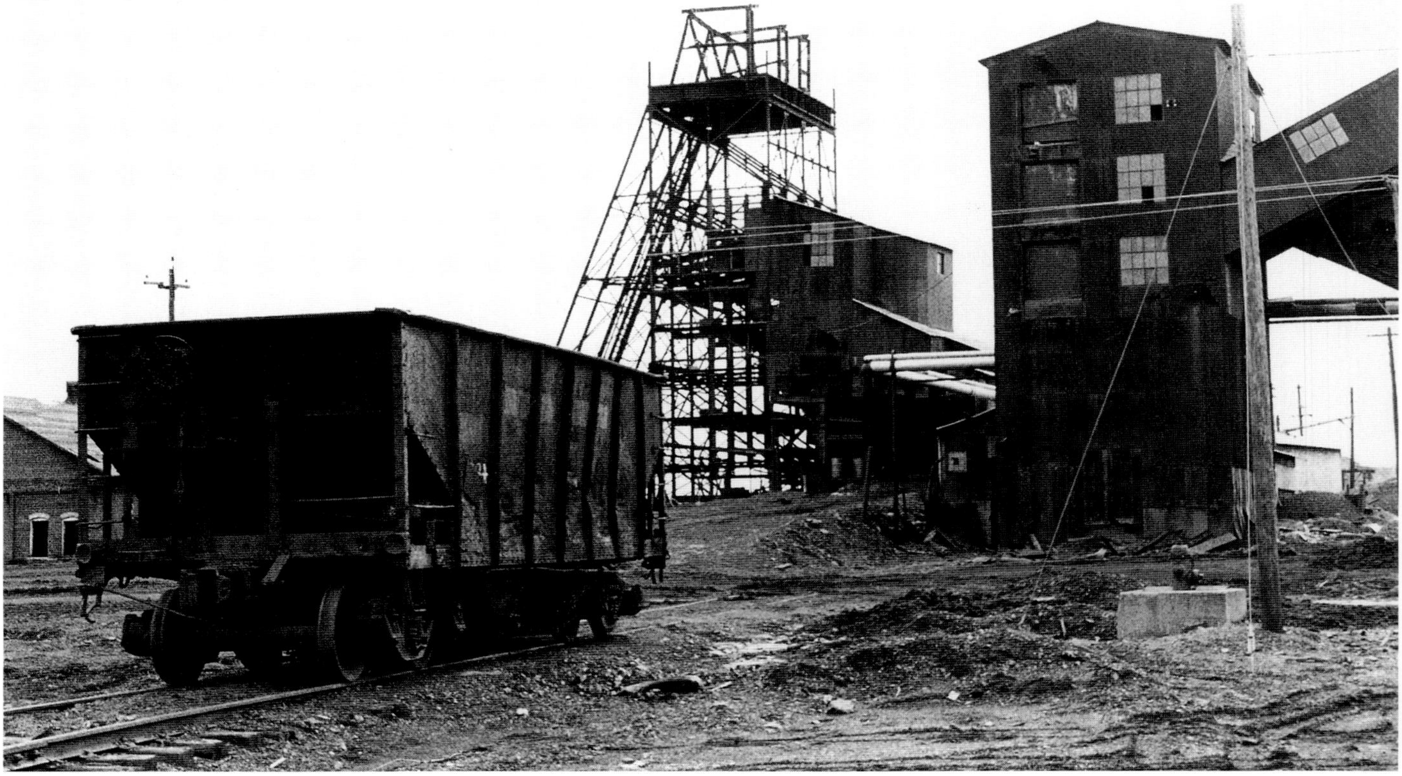

In the study of anthracite conditions, one cannot but be struck by the thought that a commodity so valuable and indispensable, lying within a small area, limited in quantity, should not be wastefully mined, and that the needs of future generations should be considered and their interests conserved.—Anthracite Coal Strike Commission, *Report to the President on the Anthracite Coal Strike May–October,* 1902

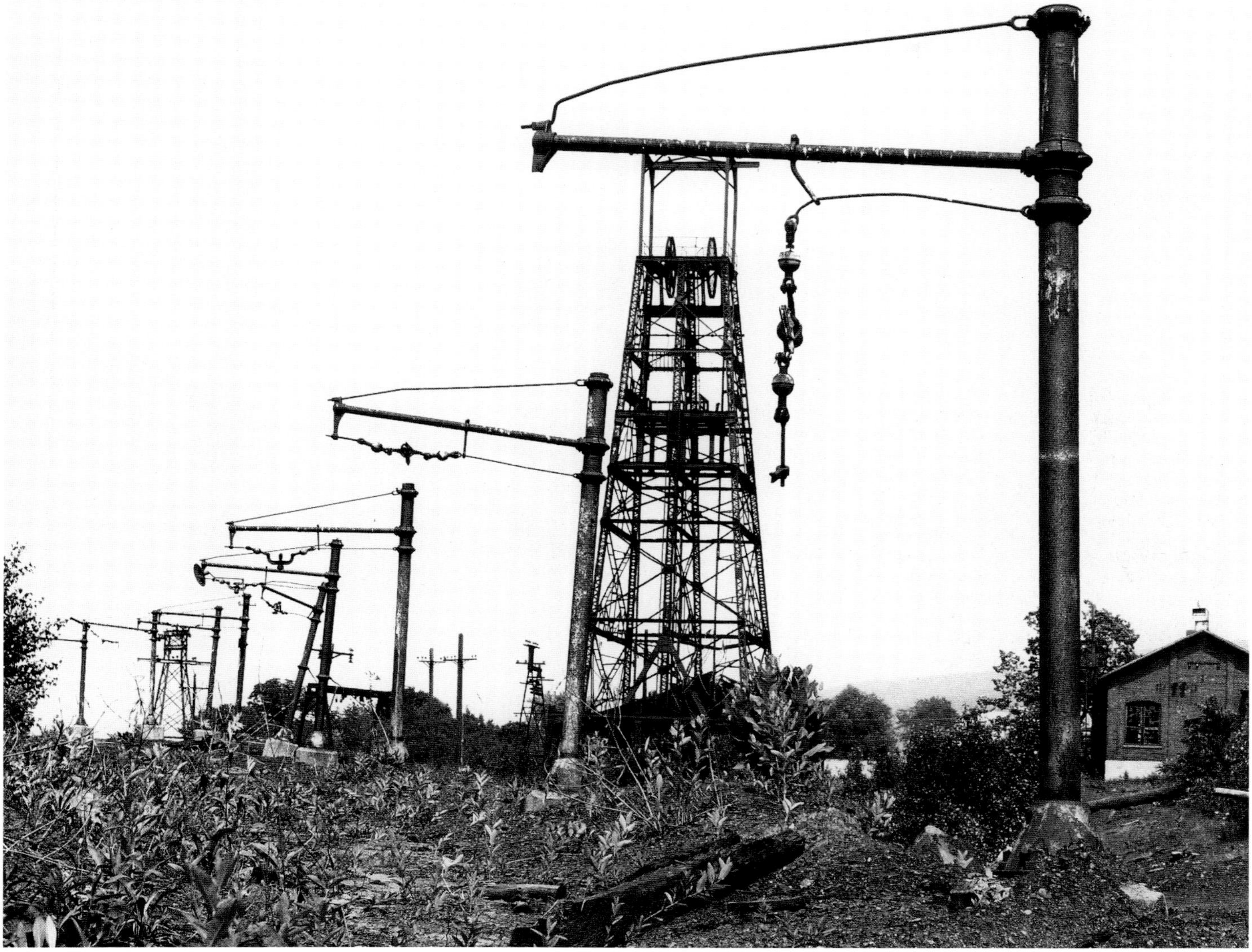

The oil and gas industries should be so regulated that coal would bear the burden of all the heavy uses and the more limited oil and gas reserves would be used only for those things which coal is not competitive.—Glen Parker, *The Coal Industry,* 1940

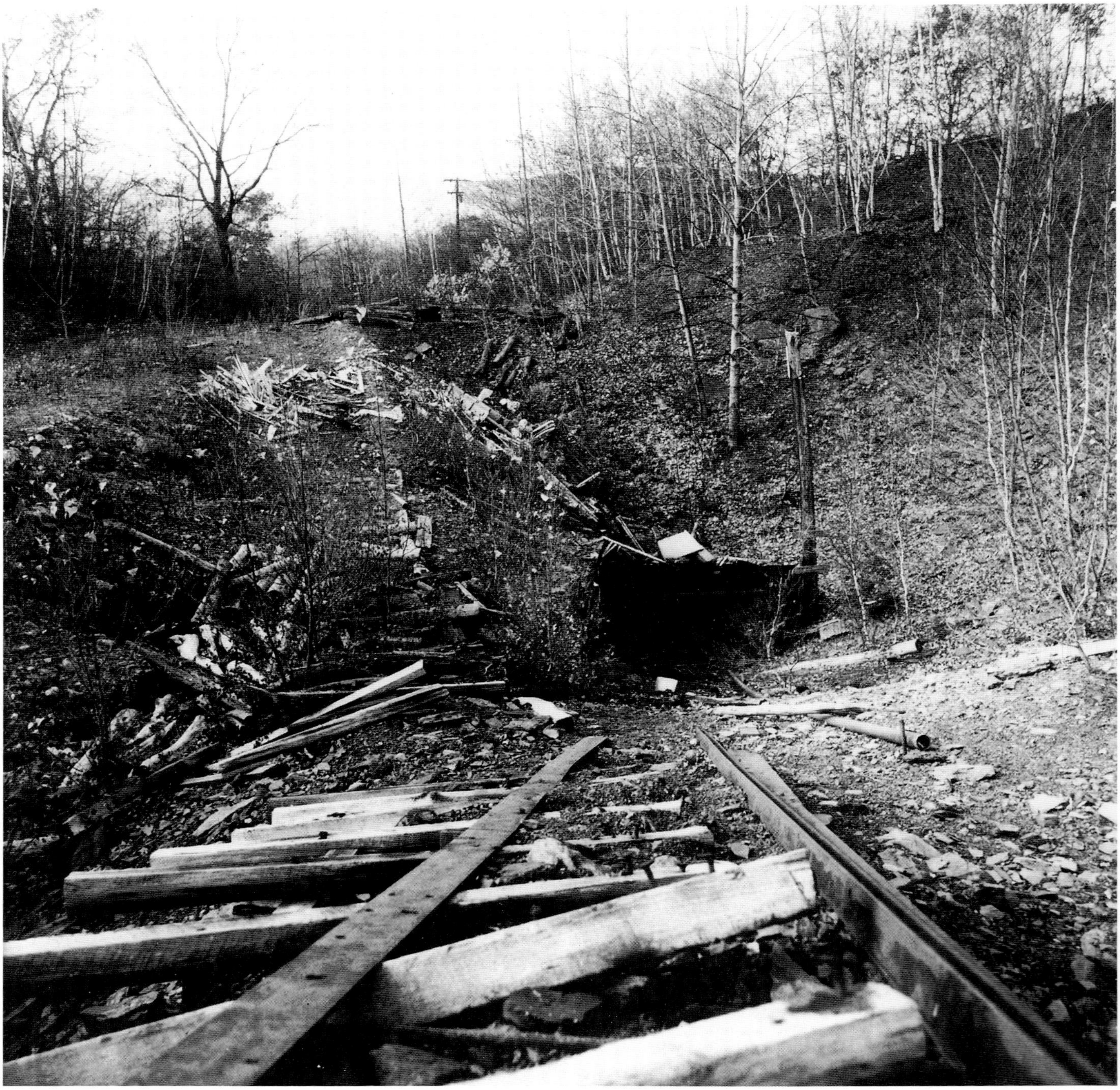

A collapsed mine.

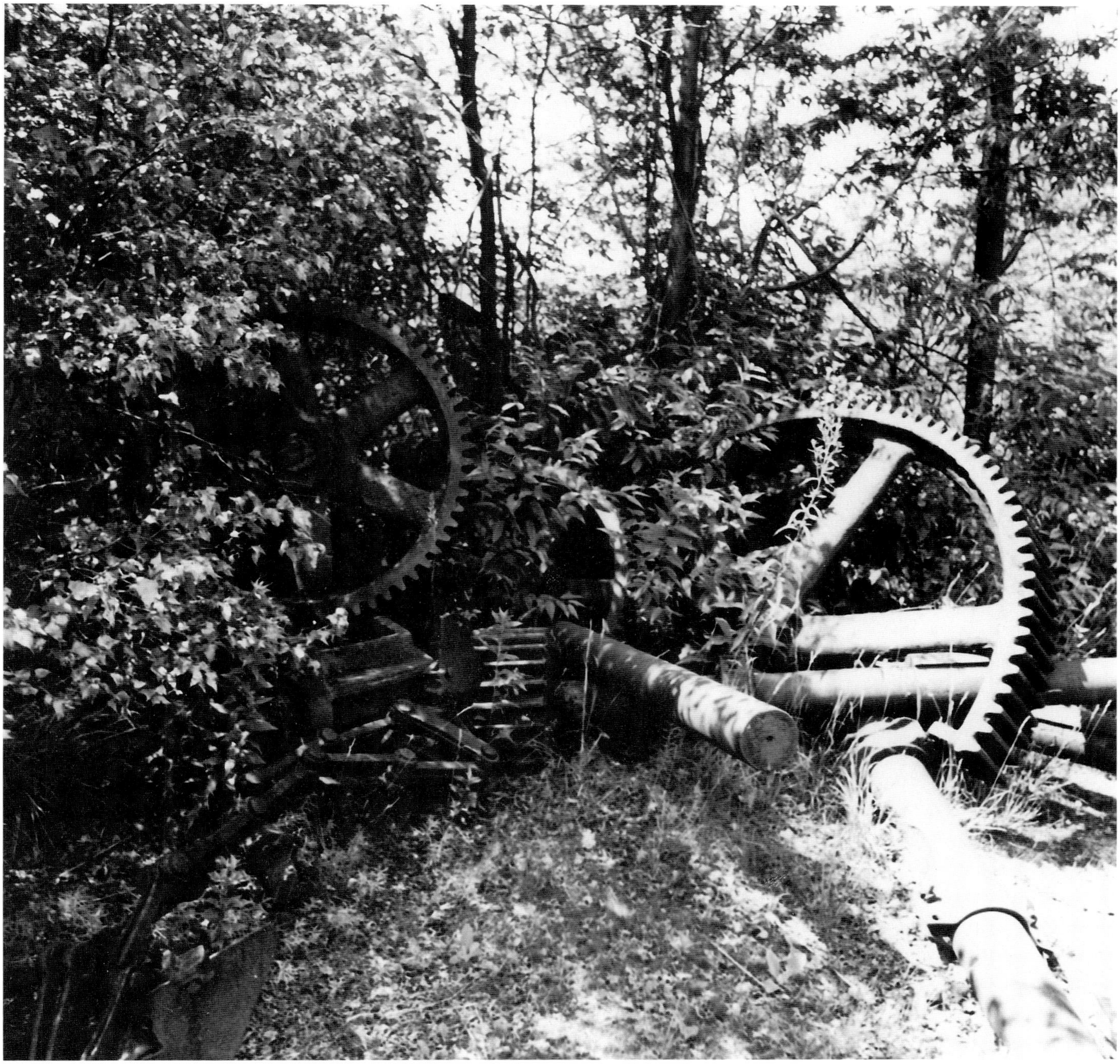

The forces of nature are slowly covering the last traces of former collieries.

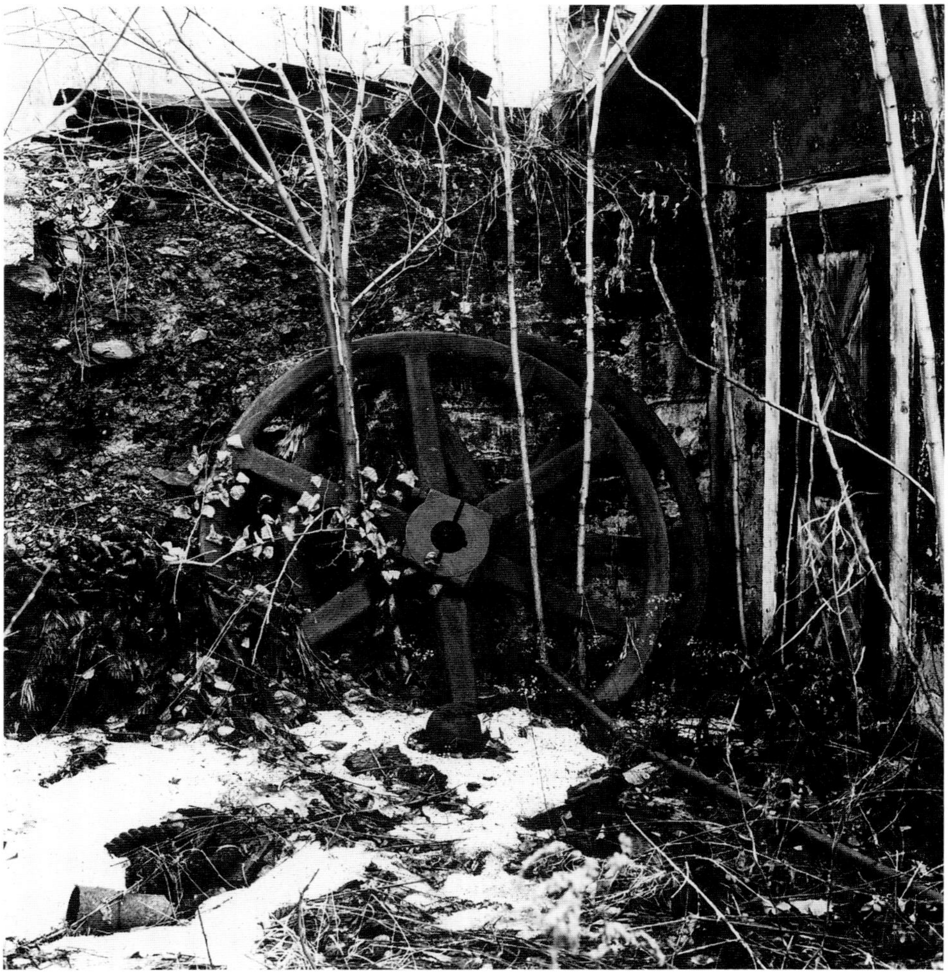

Melting snow reveals a giant wheel among breaker machinery.

In 1906, it was estimated that about 5,000,000,000 tons of anthracite coal still remained in the Pennsylvania fields. At the rate of production in 1906, this amount would last less than one hundred years.

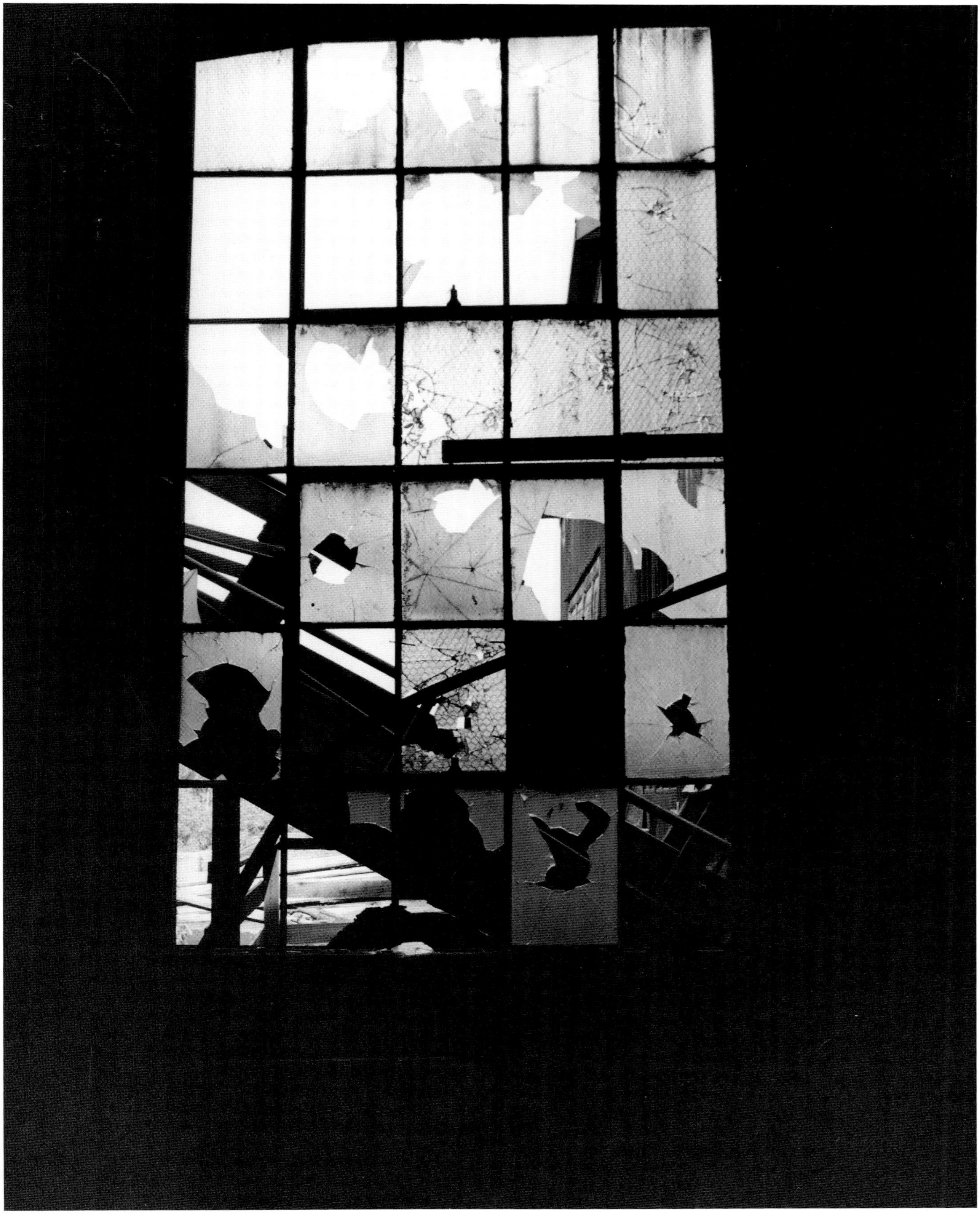

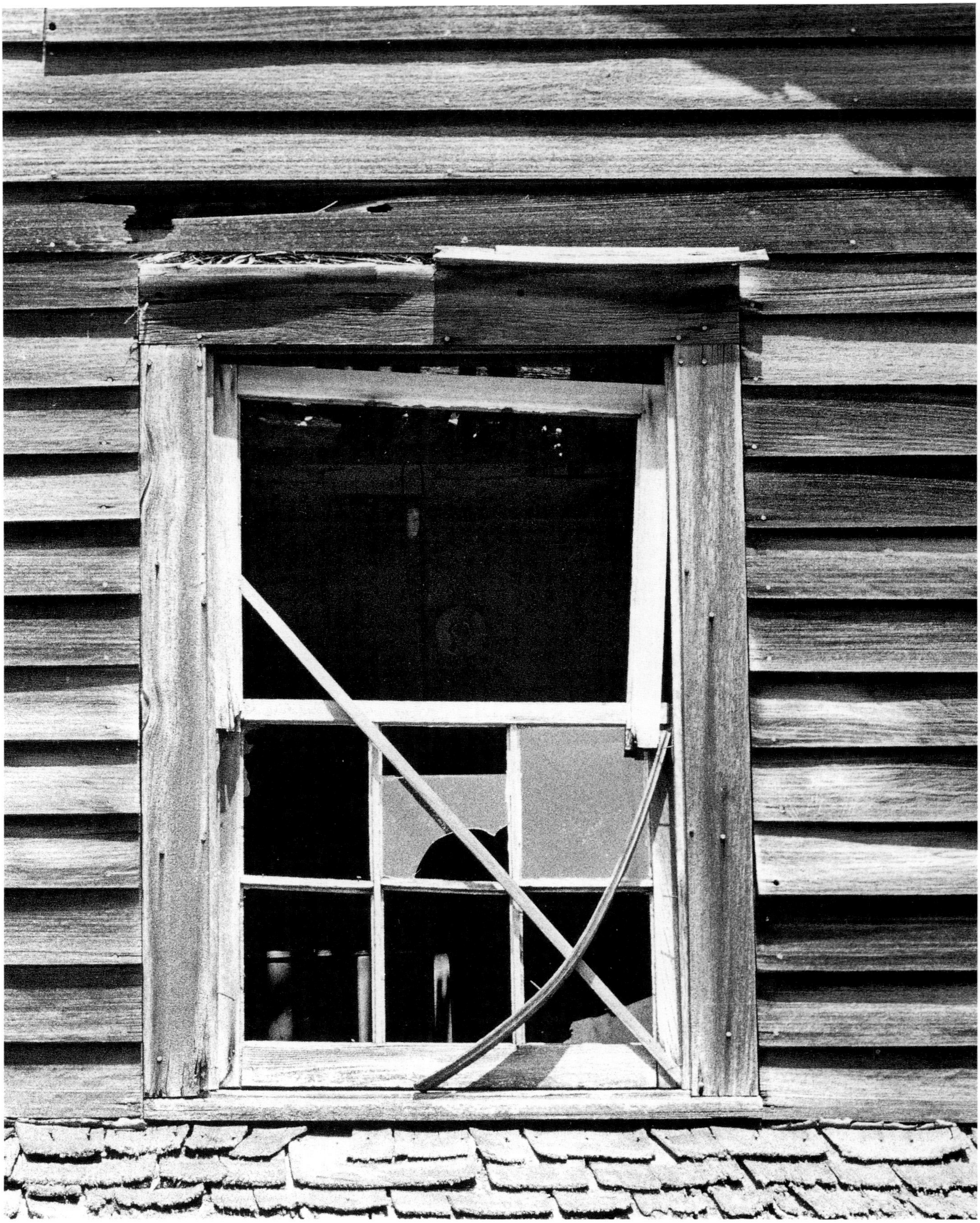

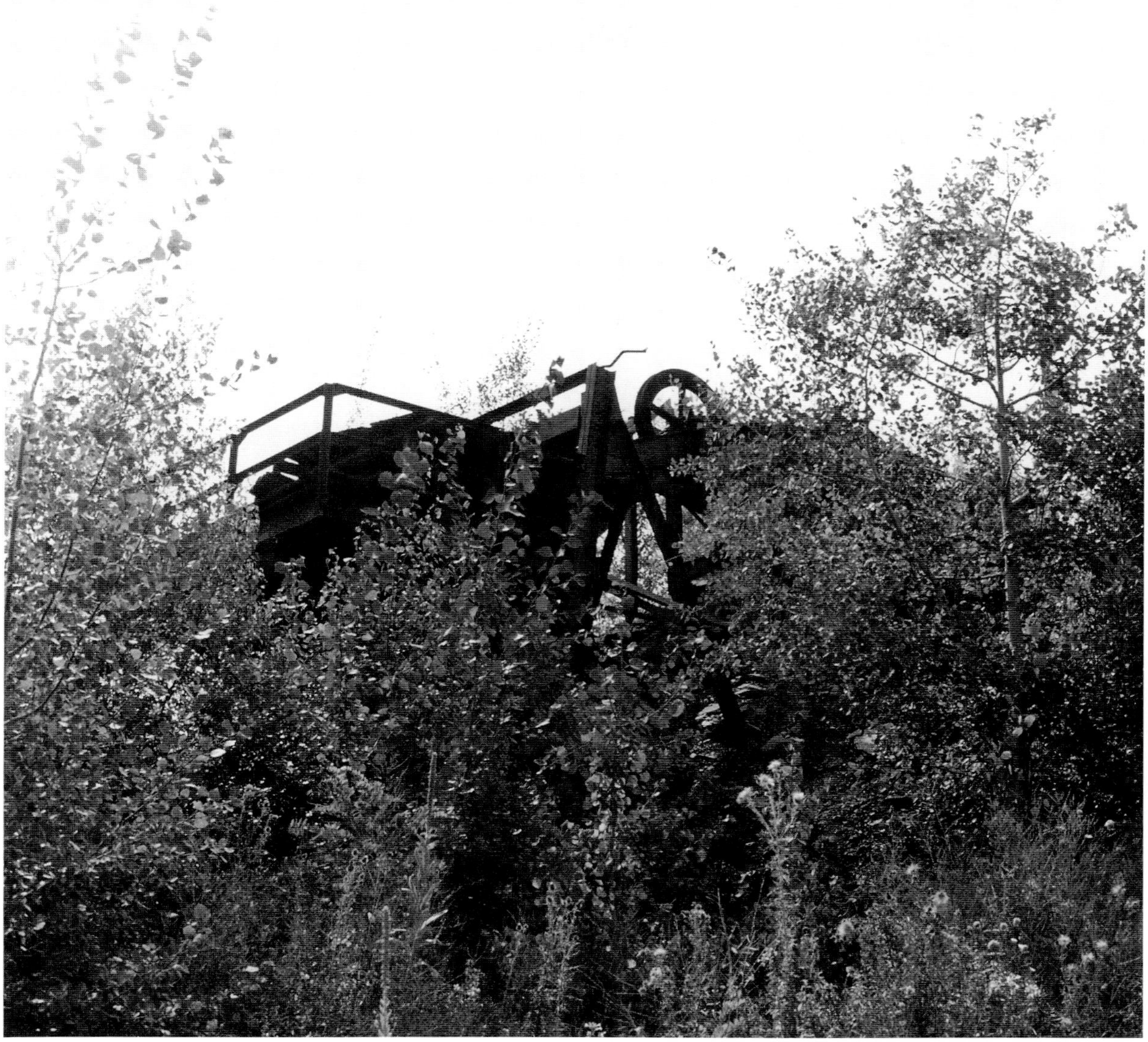

The hard lines of a tipple contrast with the trees that are gradually engulfing it.

This world is destroyed in one part, but it is renewed in another, and the operations by which this world is renewed are as evident to the scientific eye as are those in which it is necessarily destroyed.
—Jasper Williams, *Coal Catechism,* 1906

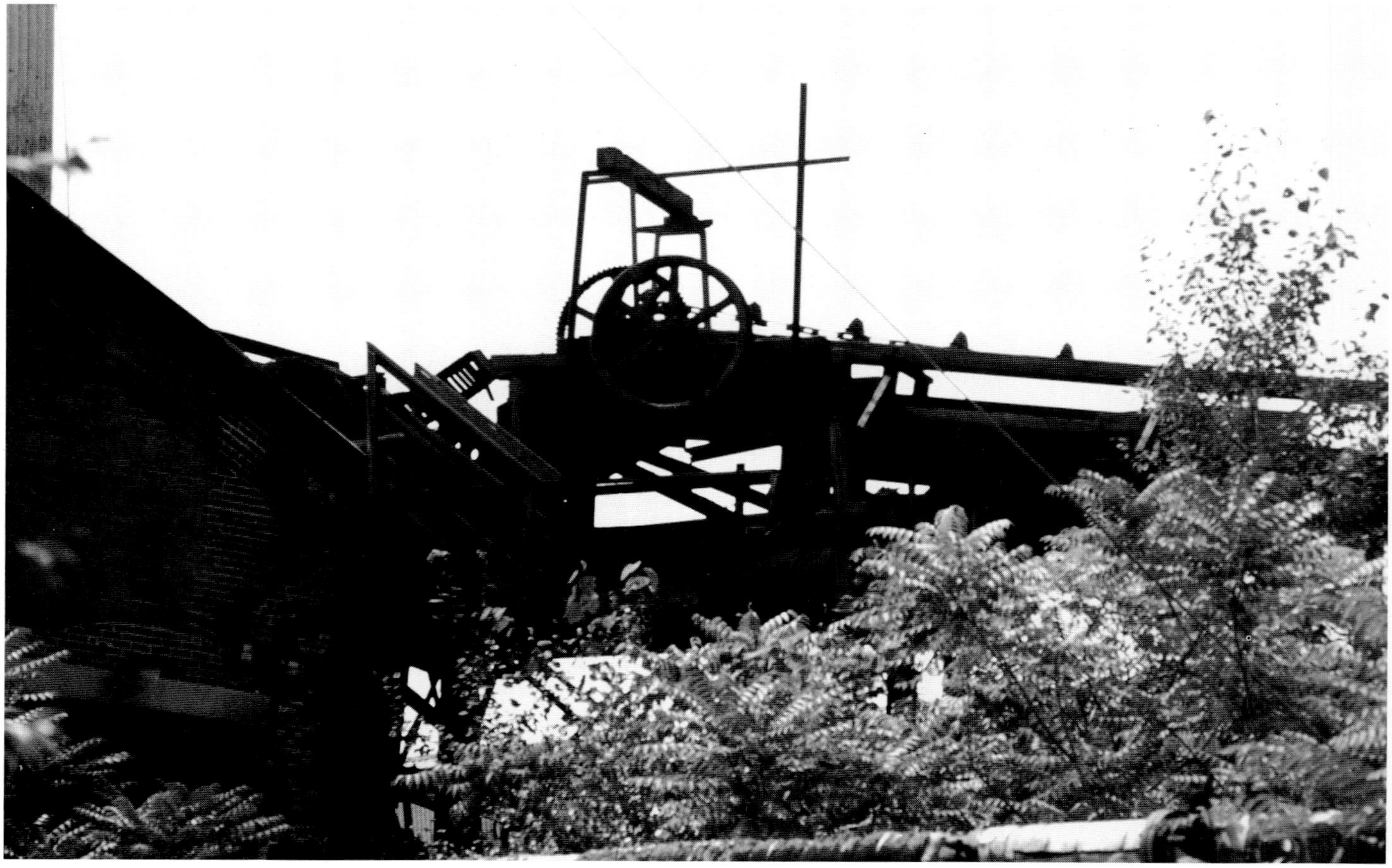

A coal tipple.

Sometimes I think that I would like to go back and see once more the big black tipple that guards the shaft-mouth and perhaps go down to B entry and watch the trains come in, and then I think of faces I would look for, faces that would not be there.
—Joseph Husband, *A Year in a Coal Mine,* 1911

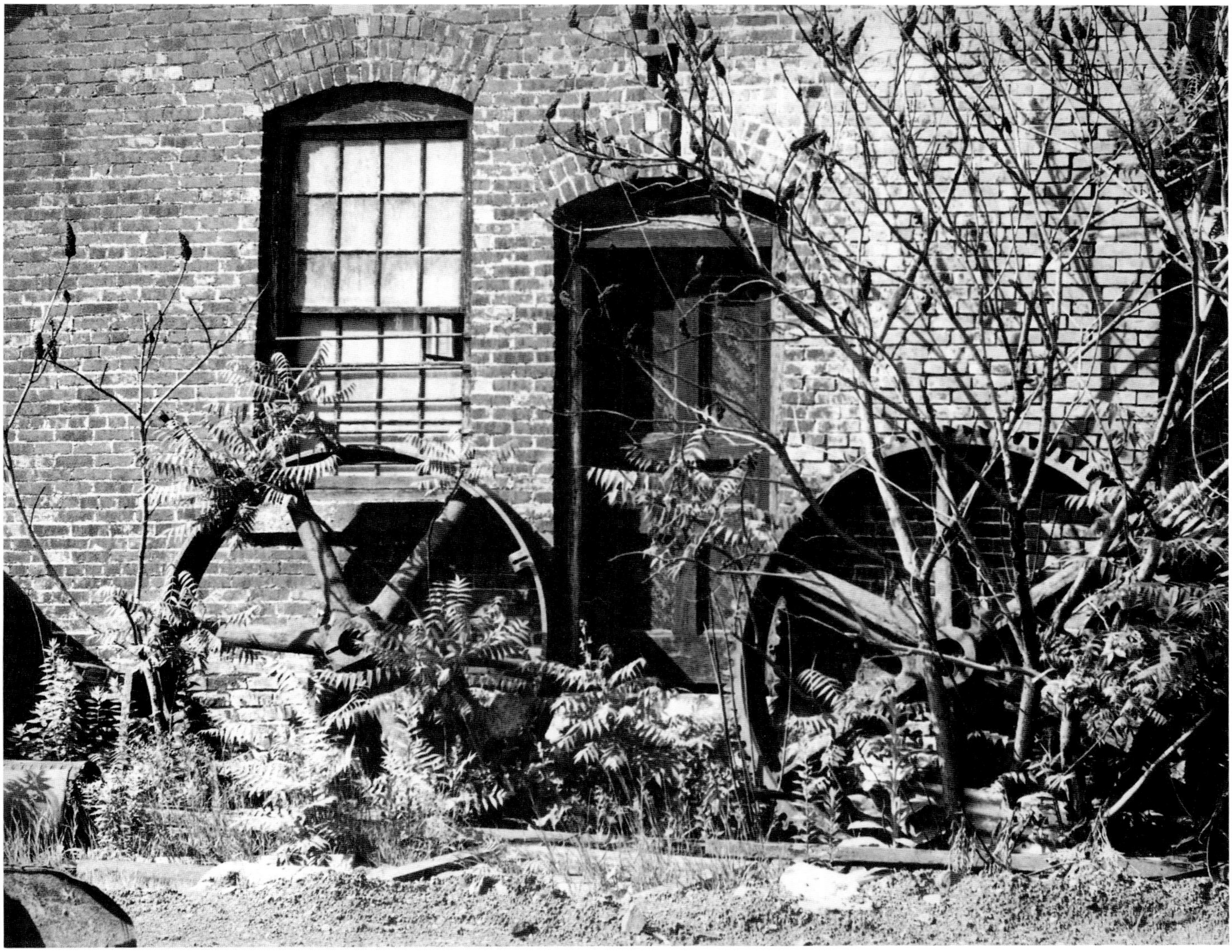

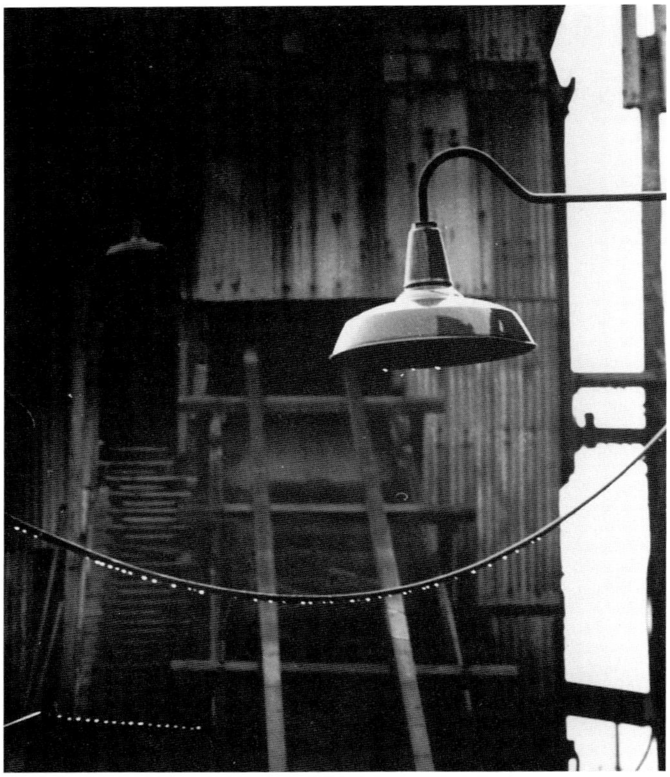

It makes me sad and lonely now
When I think of years gone by,
That I spent at the breaker
When I was a boy.
My youthful days are past and gone,
Old age is coming fast.
The only thing that is left to me
Is a picture of the past.
Years have brought on changes
To that dear little town;
The slopes were drowned out
And the breaker was torn down.
The grass is growing fresh and green
O'er the place where I used to play,
And the people who were living there.
—John J. Johnson, "Bear Ridge,
A Deserted Village," 1890

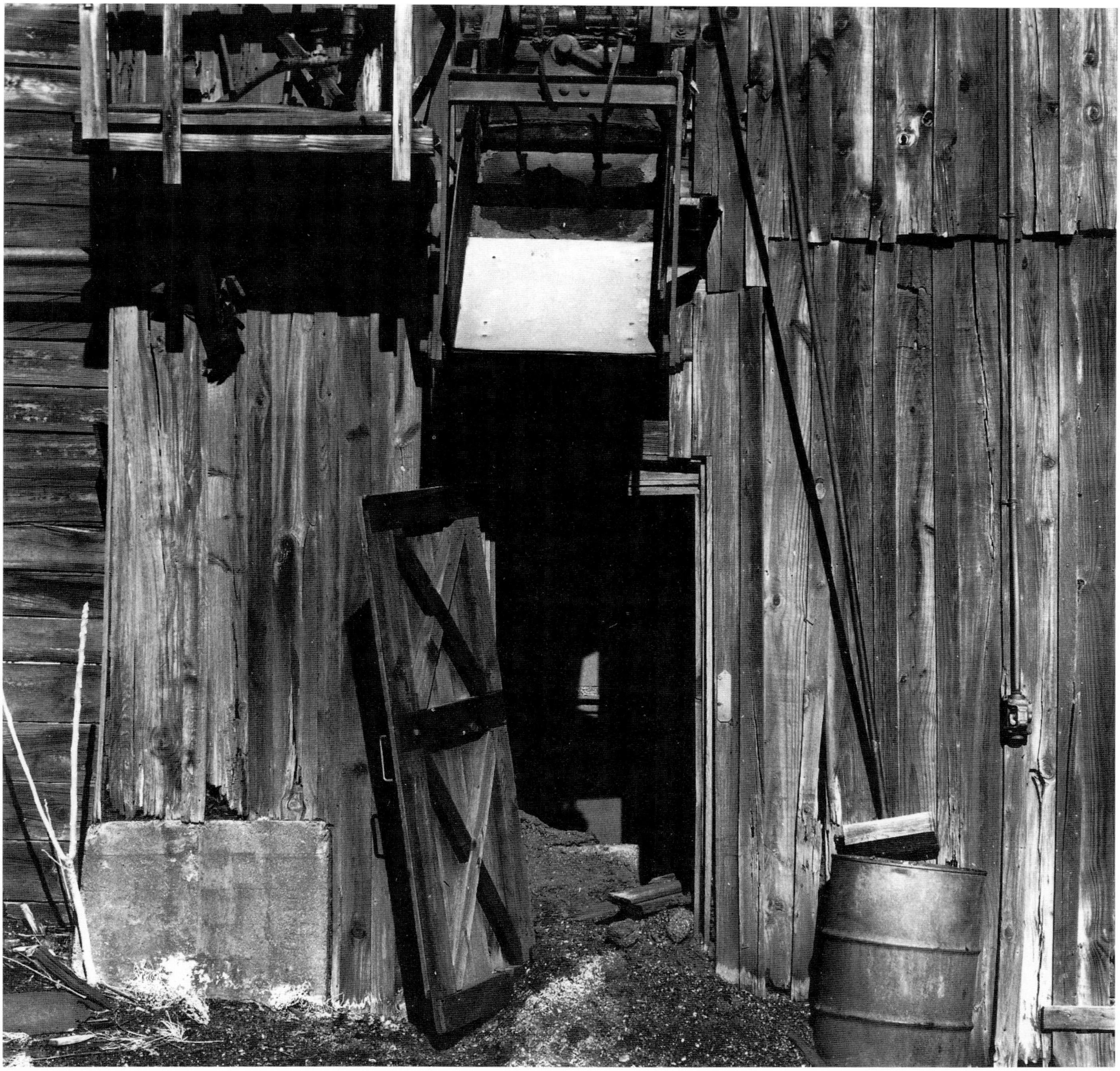

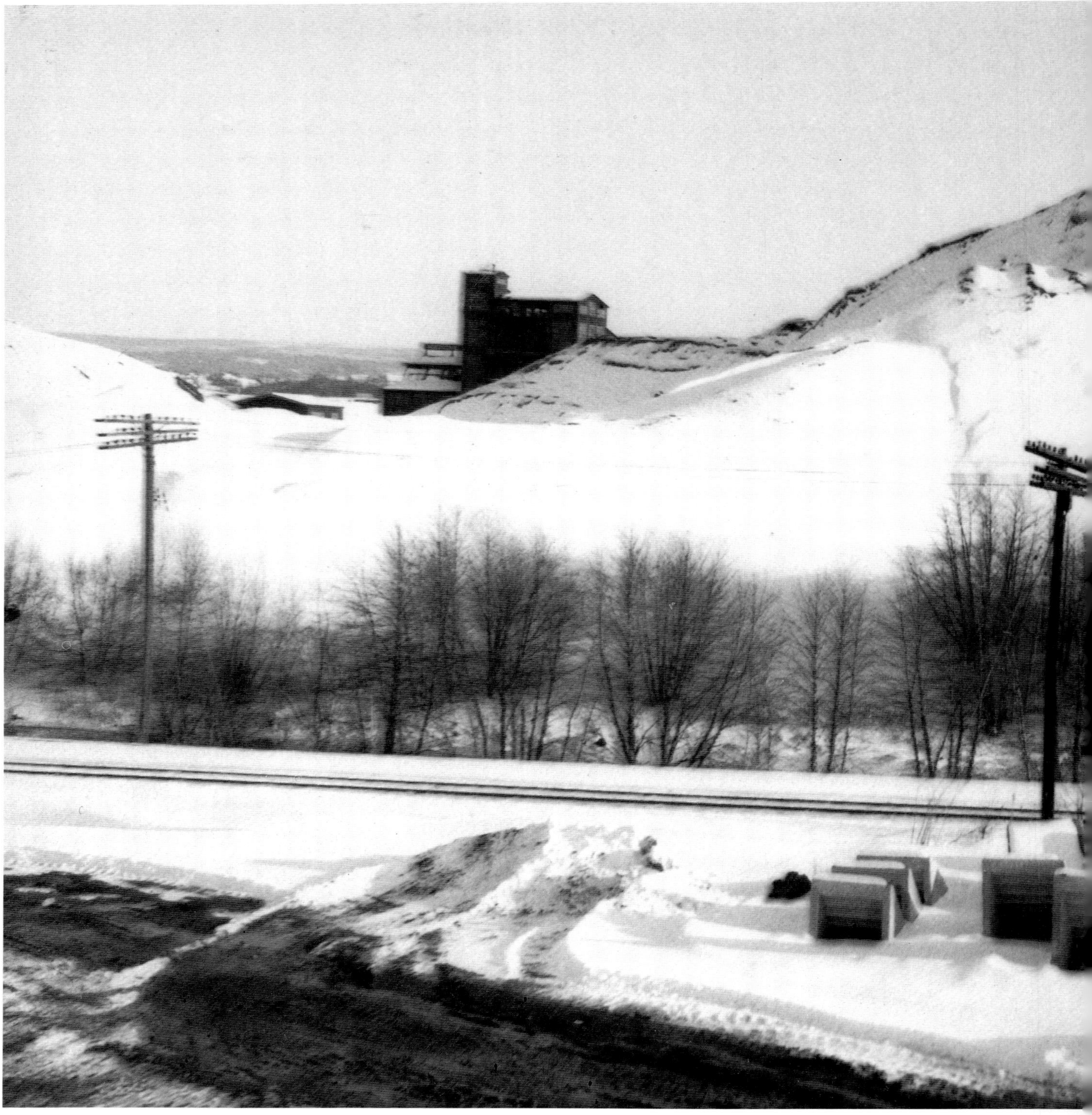

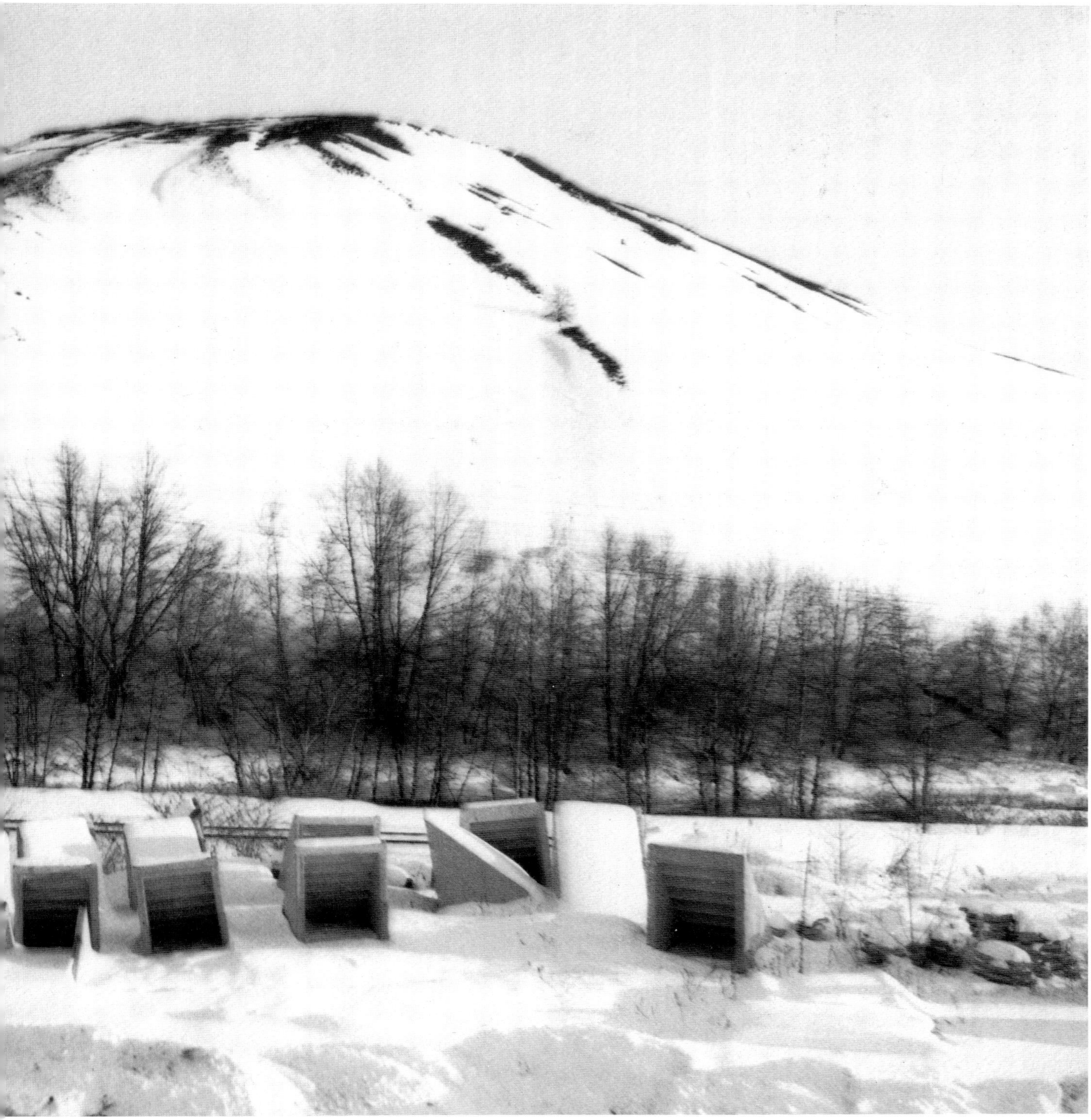

I wandered over the snow-covered coal site. The snow had melted on parts of the culm bank, revealing a thin black outline of the mountainous dump. In the distance a breaker formed a black silhouette in this scene reminiscent of a Japanese ink brush painting. As I paused, I could hear the soft murmur of the ghosts—ghosts of collieries, breakers, tipples, culm banks, and of miners whose bodies are entombed in long lost mines.—The author

Bibliography

Becker, Bernd and Hilla. *Pennsylvania Coal Mine Tipples.* New York: Dia Center for the Arts, n.d. (1991).

Blomaine, Karen (ed). *Coalseam: Poems From the Anthracite Region.* Scranton, Pa.: University of Scranton, 1993.

Bowen, George W. *Diamonds of the Mines.* Scranton, Pa.: International Textbook Press, 1928.

Carawan, Guy and Candie. *Voices From the Mountains.* New York: Alfred A. Knopf, 1975.

Federoff, D. R. "Coal, Pennsylvania Anthracite," *Bureau of Mines Minerals Yearbook.* Washington: U.S. Department of the Interior, 1970.

Gilfillan, Laura. *I Went to Pit College.* New York: Viking, 1934.

Gillespie, Angus K. *Folklorist of the Coal Fields, George Korson's Life and Works.* University Park: Pennsylvania State University Press, 1980.

Greene, Homer. *The Blind Brother: A Story of the Pennsylvania Coal Mines.* New York: Thomas Y. Crowell, 1887.

Husband, Joseph. *A Year in a Coal Mine.* Boston: Houghton Mifflin, 1911.

Korson, G. G. *Minstrels of the Mine Patch.* Hatboro, Pa.: Folklore Associates, 1964.

Lantz, Herman R. *People of Coal Town.* New York: Columbia University Press, 1958.

Lewis, Arthur H. *Lament for the Molly Maguires.* New York: Harcourt, Brace, 1964.

Mooney, Fred. *Struggle in the Coal Fields. The Autobiography of Fred Mooney.* Ed. by J. W. Hess. Morgantown: West Virginia University Press, 1967.

Morris, Nelson. *Scavicky: Portrait of an Anthracite Family.* Wilkes-Barre, Pa.: Wilkes University Press, 1992.

Nearing, Scott. *Anthracite: An Instance of Natural Resources Monopoly.* Plainview, N.Y.: Books for Libraries, 1915.

Nicols, William Jasper. *Coal Catechism.* Philadelphia: J. B. Lippincott, 1906.

Parker, Glen Lawhon. *The Coal Industry, A Study in Social Control.* Washington: American Council on Public Affairs, 1940.

Poliniak, Louis. *When Coal Was King.* Lebanon, Pa.: Applied Arts, 1970.

Roberts, Ellis. *The Breaker Whistle Blows.* Scranton: Anthracite Museum Press, 1984.

Smith, S. P. *The Black Trail of Anthracite.* Kingston, Pa.: S. P. Smith, 1907.

Thomas, Maude Morgan. *Singing in the Dark: The Story of the Welsh in Pennsylvania.* Philadelphia: John C. Winston, 1954.

Wooley, Brian. *We Be Here When the Morning Comes.* Lexington: University Press of Kentucky, 1974.

Acknowledgments

Grateful acknowledgment is made to the following for permission to reprint previously published material:

Guy & Candie Carawan: for excerpts from *Voices From the Mountains.* Copyright © 1975 by G. & C. Carawan.

Harcourt, Brace & World: for an excerpt from *Lament for the Molly Maguires.* Copyright © 1964 and renewed 1992 by Arthur H. Lewis.

International Publishers, Inc.: for excerpts from *The Molly Maguires* by Anthony Bimba.

Penguin Books USA Inc.: for excerpts from *I Went to Pit College* by Laura Gilfillan. Copyright 1934 by Laura Gilfillan, © renewed 1961 by Harriet W. Gilfillan.

University of North Carolina Press: for an excerpt from *Cloud by Day: A Story of Coal and Coke and People* by Muriel Earley Sheppard. Copyright © 1947 by the University of North Carolina Press.